Social Science and the Self

Social Science and the Self

*Personal Essays
on an Art Form*

Susan Krieger

*Rutgers University Press
New Brunswick, New Jersey*

Library of Congress Cataloging-in-Publication Data

Krieger, Susan.
Social science and the self : personal essays on an art form
Susan Krieger.
p.   cm.
Includes bibliographical references.
ISBN 0-8135-1714-1 (cloth)—ISBN 0-8135-1715-X (pbk.)
1. Social sciences.   2. Social sciences—Methodology.   3. Social
scientists.   4. Self.   I. Title.
H35.K68   1991
300—dc20   90-28754
                                                                    CIP

British Cataloging-in-Publication information available.

For my brother Daniel Cahn

Contents

Acknowledgments   *ix*

Introduction   *1*

I
A Social Scientific Education

1   Self and Context   *11*

2   Teachers: Approaches to Interpretation   *16*

II
Forms of Expression

3   The Vulnerability of a Writer   *29*

4   The Presence of the Self   *43*

5   Speaking of Writing   *49*

6   An Anthropologist and a Mystery Writer   *56*

Contents

III
Individuality

7 Self, Truth, and Form: Lessons from Georgia
   O'Keeffe   67

8 From O'Keeffe to Pueblo Potters   85

9 Pueblo Indian Potters: Individual Difference in a
   Collective Tradition   91

10 Psychotherapy and Pottery Making: Approaches to
    Self-Knowledge   121

IV
Teaching and Research

11 Experiences in Teaching: Exposure, Invisibility,
    and Writing Personally   135

12 Snapshots of Research   150

13 Beyond Subjectivity   165

V
Other Voices

14 Problems of Self and Form, I   187

15 Problems of Self and Form, II   216

Notes   245

Bibliography   265

Acknowledgments

THIS BOOK HAS TAKEN LONGER to write than I expected. My first attempt to put some version of it into words was funded by the Russell Sage Foundation during 1984–86. I am grateful to Marshall Robinson for his encouragement at that time. The present volume was drafted during 1988–90. I would not have undertaken it without the interest in my work shown by Kenneth Arnold of Rutgers University Press, who was willing to offer institutional support during a period when I was feeling unusually discouraged. He said he was interested in publishing something radical, which allowed me to do what I most wanted, a book on personal involvement in social science.

In the process of writing the book, people from different fields helped me with their comments on the manuscript and gave me positive feedback when I needed reassurance that my work and, by implication, myself would be appreciated by others. I want to thank Peter Adler, Marjorie DeVault, Kathryn Gohl, Patricia Gumport, Martin Krieger, James G. March, John Van Maanen, the students in the feminist methodology course I taught at Stanford in spring 1990, and those of the women I interviewed for the "Other Voices" section of the book who subsequently commented on the entire manuscript. Ann Swidler's advice on the earlier manuscript continued to be helpful as I worked on this later version.

Many people gave of themselves as participants in the research projects I have drawn on in this book. The women interviewed for the "Other Voices" section were extremely generous in allowing me to use their interview material as I wished. They also were friendly and enthusiastic about the subject of the book during our interview sessions, and this gave me encouragement to continue. The students whose papers and classes I mention in the chapters on teaching gave of themselves by sharing their experiences honestly with me. I continue to be indebted to the people who participated in the prior studies I reflect upon in the discussions of research and writing. The experiences of people I have studied stay with me long after any research is completed.

Personal losses can affect a study as much as deliberate research strategies. This book was written during the AIDS crisis of the 1980s. Living in San Francisco, I could not help but be affected by the presence of so many gay men who would be dying. During this period, my younger brother, to whom this book is dedicated, also died, a probable suicide. These tragic events made me want, all the more, to register the importance of the individual human life.

The value of my own life, when I doubt it, is made clear to me by two women whose care and support I want to acknowledge. Carolyn Hallowell was my psychotherapist while I was working on this book. Many of the themes I discuss in these pages have also been the subject of our therapy sessions. I am grateful to Carolyn for emphasizing the desirability of self-expression, for her emotional generosity, and, most of all, for her willingness to respond to my need for her. Without her, I might still have written this book, but I would not have done it as well.

Estelle Freedman is the woman with whom I share my life most intimately. She has been the nicest surprise of each day for the past ten years. Estelle has read every page of this book many times and has changed it in more places than I would like to remember. While I was writing the manuscript, she would repeatedly tell me to spell out my meaning for my readers, which had the effect of forcing me to come to know what my main points were. To write a book like this takes more self-confidence than I alone have. I would not be discussing my own experiences to the extent that I do in these pages if Estelle had not encouraged me. The individuality that I speak of as so vital to social scientific insight is really, in my experience, a product of joint efforts; nowhere have I been more aware of joint effort and shared joy than in my relationship with Estelle.

Social Science and the Self

Introduction

THE SOCIAL SCIENCE DISCIPLINES tend to view the self of the social scientific observer as a contaminant. The self—the unique inner life of the observer—is treated as something to be separated out, neutralized, minimized, standardized, and controlled. At the same time, the observer is expected to use the self to the end of understanding the world. My central argument in this book is that the contaminant view of the self is something we ought to alter. I think we ought to develop our different individual perspectives more fully in social science, and we ought to acknowledge, more honestly than we do, the extent to which our studies are reflections of our inner lives.

The social sciences, however, are premised on a type of thinking that limits discussion of the self. We are taught to avoid attention to the authorial first person, whose view, and whose choices, a study represents. We learn to become invisible authors. If we cannot be objective, at least we should not call too much attention to the fact of our subjectivity. We learn to speak in a standard way, rather than in terms that might be revealed as our own. The conduct of science is said to depend on the development of consensual knowledge, or on a common view. The idiosyncratic, individual view, and the basis for that view in an observer's inner life, is often considered unimportant, irrelevant, or wrong.

I argue, in this book, for the inner, individual view, and for the importance of developing it in our studies. I do not think that the more full development of individual and inner perspectives will result in the downfall of social science, nor will it lessen our abilities to understand the world outside ourselves. Rather, I believe that increased personal understanding can help us think more intelligently and fully about social life.

My argument has its basis in my own experience. In recent years, I have increasingly come to think in personal terms about the social science I do. At one time, I could write a study and then write separately about how, and why, I came to do it. I no longer feel I can proceed in that way. I now need to speak more directly about my involvement with any subject I study. Writing about others, or about a social process, without reference to the self has come to feel alienating and untrue to me. Writing personally has become a way that I can feel I am doing social science in a responsible manner. In speaking of myself in my work, however, I have come up against barriers. These are ideas I have internalized at least since graduate school. They reflect the fact that social science ideology is full of prohibitions against conspicuous use of the self. It is full of guidelines that suggest that the standard forms of expression are the only correct forms.

This book is about a struggle with the standard forms and with traditional styles of expression within social science. It is about a wish for something better, for a social science that does not deny the self, but that seeks to use its potential. It is about a social science that, in the long run, might be more rewarding for many of us than the limited version of science that we learn about in most methodology texts. The science I speak of is not hard, objective, standard, dispassionate; nor is it about measurement, data, clear-cut models of behavior, or procedures for testing. It is soft, subjective, idiosyncratic, ambivalent, conflicted, about the inner life, and about experiences that cannot be measured, tested, or fully shared.

Although the term "art" is in my title, I do not argue, in these pages, that social science ought to be more like art. Nor do I argue that social science is part science and part art, although it may well be. Rather, I use examples from the world of art to speak of values and ways of thinking that usually are not part of how we think about social science.[1] I discuss the relationship between self and work that is described in the writings of Georgia O'Keeffe, for example, and in statements made by Pueblo Indian potters. What are the implications for social science? I ask.

I begin my more general discussion by speaking specifically about myself and how I was trained as a social scientist (part I, "A Social Scientific Education"). I then discuss both personal and general issues that arise in the writing of social science. Next, I branch out to explore ideas about authorial perspectives held by a mystery writer and an anthropologist; ideas about self, truth, and form in the writings of O'Keeffe; and ideas about individuality and conformity in statements made by Pueblo Indian potters (parts II and III, "Forms of Expression" and "Individuality"). I subsequently return to myself for a discussion of teaching and research experiences in which I continue to deal with themes raised in earlier chapters (part IV, "Teaching and Research"). The book closes with a discussion that further elaborates on these themes, but this time the voices are those of eight other feminist scholars reflecting on the relationship between self and form in their studies (part V, "Other Voices").

Throughout this volume, I focus repeatedly on the subject of writing, for written work is the central way I have come to understand what social science means. In my sociological writing, and in my personal autobiographical writing, I have been concerned with the problem of how to depict particular social realities: how to get at what is true, or important, and how to see, in specific situations, dynamics that are more broadly meaningful. I have written two earlier sociological case studies that have concerns similar to this one. The first was a study of a rock music radio station (*Hip Capitalism*), the second a study of a lesbian community (*The Mirror Dance: Identity in a Women's Community*).[2] Although the settings were different, each of these studies focused on problems of the individual sense of self, and on difficulties individuals experienced in relating to an established order. More important, perhaps, both these studies were written in an unusual fashion. They described social situations concretely and specifically, without the theoretical language that is common in social science. In both studies, I relied primarily on the language of people I interviewed, and I used their language to depict dynamics of their situations. I thought that a sociological account could speak very powerfully through its details, and that one could structure a detailed description to provide an example that said a great deal about a larger social problem.

In the present study, I continue in the methodological tradition of speaking specifically in order to speak generally. However, in this work I use the specific case of my own experience as my main example. I extend this example by elaborating on the experiences of others: O'Keeffe, the potters, and other academics. A self-other counterpoint can be felt throughout this volume. However, for me, the relationship between myself and others is largely one of continuity. I draw on discussions by others in a similar fashion as I draw on my own discussion. I look for how the persons I quote view themselves in relation to their work. I look for how they are part of the visions they create. I find themes embedded in their experiences that are similar to those I find in my own experience, themes such as a fear of exposure joined with a fear of invisibility, and a struggle to assert individuality despite pressures toward conformity. I focus on details to articulate my themes, feeling that the specifics are more important than any generalizations I can make. Nonetheless, I do make some generalizations so that my particular points will be clear. These normally occur at the ends of subsections and chapters.

In writing this book, I have pushed myself to speak both more personally and more generally than I thought necessary at first. I have often said more about an experience of my own than came easily to me. And I have pushed to say more of a generalizing, and simplifying, nature than I tend to believe I should have to say. For me, the strongest statement is always the direct personal one. That type of statement brings me closer to an experience and helps me feel grounded and honest. It gives me more confidence concerning what I say than a general statement does. Readers may find, however, that my discussion could be further developed in either a personal or general direction. I want to note only that my remarks are already developed, in both respects, far beyond what I would normally expect of myself.

Because much of this account is written in a personal style, I should mention that personal statements often affect readers in different ways than do more abstract conceptual statements. The effects of personal statements may not be comprehended immediately, and they may never be comprehended in strictly intellectual terms. For instance, one's first response to a statement describing a personal experience might be to reject it, or to find it lacking; or it might be to feel that the statement explains a great deal, or is telling. In either case, one may later feel the reverse, or feel differently, or one may draw different lessons from the same experience. My discussion, then, is offered as material to reflect upon over time. I wish it to be a statement that works emotionally as well as intellectually.

On both levels, my central argument in this book is that when we discuss others, we are always talking about ourselves. Our images of "them" are images of "us." Our theories of how "they" act and what "they" are like, are, first of all, theories about ourselves: who we are, how we act, and what we are like. This self-reflective nature of our statements is something we can never avoid. In social science, although we try to comprehend others, and although we may aim to depict the ways their realities are different from our own, understanding others actually requires us to project a great deal of ourselves onto others, and onto the world at large. It also requires taking others into the self in an encompassing way. Although I do not use psychoanalytic terms in this book, my thinking will not be foreign to depth psychologists, who are used to talking in terms of object relations, introjections, projections, and selfobjects. Social scientists, in general, however, do not talk in a self-centered and internal way. We tend to believe that the world outside the self is the world of primary importance. The external world is the world our studies ought to be about. We may concede that the external world is known through us, and that a considerable amount of empathy, identification, or observer involvement is required for social knowledge, but it is usually not said that all, or most, of what we know is ourselves.

I am saying that I feel it is. For me, it makes more sense to see the world as self than to imagine that we can know a free-standing external reality, or to say that what we know is a largely unweighted interaction between self and other. A reader need not subscribe to a belief that puts such a great emphasis on the self in order to find my discussion helpful though. One need only be interested in acknowledging the self more fully in social science. My position of seeing the world in terms of the self bears noting primarily because it informs this work. It is the position that has seemed strongest for my purpose of developing an alternative to traditional social science views about the self. To me, seeing social science as an expression of the self means that I cannot make a judgment about whether an author, or a work, talks too much about self and not enough about other, or not enough about an external world. Rather, I have to ask if the author, or work, talks about self, and world, in a way I find interesting, useful, or valuable.

In my view, there is no right balance between self and other in a study. There are simply different ways of expressing, or using, the self. Similarly, it is not the match of someone else's sense of reality to the world that I can judge, but the response that person's sense of reality evokes in me. If someone else's sense of reality fails to interest me, I have to see it not as that person's failure, but as mine. I have to see it as a failure in my ability to respond. I also have to see all attempts at talking directly about the self as having value. I think that only when we begin to speak of ourselves in social science explicitly, and at greater length, will we be able to speak of ourselves better. It is important to reveal not only more of the outer world on which we focus our gaze, but more about the inner worlds in which we assemble what we choose to say.

I do not claim to have achieved a final best solution to the problem of use of the self in social science, or even to the problem of use of my own first-person voice. My social science writing is affected by the many constraints I feel. These have sources in social science ideology, in the culture in general, and in my own personal history and ability. Nonetheless, despite my limitations, I would like the spirit of my work to come through. I hope the reader will find significant themes in the various accounts that compose this book. I would like the book to be a stimulus for further thought and feeling, and I hope it will be read along with the question, What are the implications for my own work?

This book is intended as a contribution to social science method and to the field of feminist scholarship. I was trained as a sociologist and broadly influenced by the methodological ideology of the social sciences. I continue to be grateful for early encouragement in the subfield of organizational sociology, and I place myself most easily in the fieldwork or ethnographic tradition. In recent years, I have come to see value in identifying my work as feminist scholarship as well. It is in the area of feminist inquiry, within the humanities and the social sciences, that I have found the most recent explicit commitment to many of the values that are central to my work and that make it deviant from the standard work in social science. I have written this book to be of use to those who share many of my struggles and concerns. This is a group not limited to women, feminists, or qualitative social scientists, but I think it would be a mistake to overlook the book's primary importance to people who identify in these ways.

In discussing my experiences, and in my selfish use of the experiences of others, *Social Science and the Self* is both about its subject and a demonstration of what it advocates. It is not a resolution of "social science and self" dilemmas so much as it is an articulation of them. Thus, my use of quotations from others throughout is closely tied to the themes of my own attempt to speak. As I hear the potters, the painter, and the academics speaking, I also hear myself. I ask the reader to do the same with them and with me.

I

A Social Scientific
Education

Self and Context

As an undergraduate, I did not expect to become an academic. I majored in sociology as a route to becoming a city planner. I thought that a doctor was the thing to be, but I did not like blood so I would save cities. I became so involved in studying planning that issues of sociology as a discipline, or social science as a method, were not much on my mind. Although academic work was what my professors did, I had no sense of how they did it and no feeling that such work was admirable. I wanted to end poverty and change the way we lived.

During my two years as a graduate student in urban planning, I learned about the value of interdisciplinary work and about the social meanings of physical land use patterns. I was taught to think like a welfare economist, and I came to understand a term called "rationality," which had to do with intelligence in assessing ends and means. In planning school, I also met doctoral students for the first time and, soon after that, reoriented myself to the goal of getting an advanced academic degree. Applying for a doctorate meant, to me, that I had to internalize the idea that the pursuit of knowledge was a worthy cause and that I had to make a commitment to that pursuit. I had to adopt an ideal that was different from my prior goal of bringing about social change. The knowledge ideal felt more passive to me and more associated with wisdom. It was not unrelated to improving the world, but it felt darker as a goal, more internal, less clear.

After working as a planner for a year, I enrolled in a doctoral program in communication research, a relatively new social science. I initially thought I would become a communication policy planner and do research that would affect national media decisions. In graduate school, I became a member of a group that moved through a doctoral program together. My first-year class consisted of twelve students who would each take five or more years to complete the program, although some initially boasted they would do it in three. The most influential of our classes was a theory seminar that met each week throughout the first year. In this course, our graduate class got to know one another. We learned the basic concepts of the field, which was defined broadly as covering all human communication, and we met the different individual faculty members in our department. Each faculty member made a guest appearance in the seminar, while one member regularly ran it. In the theory seminar, more than any other, I began to see academic socialization in action.

In addition to content, and more important than learning the history of the field, it was evident that we were learning to reproduce a style. This style was displayed most prominently by the faculty member who regularly conducted the theory seminar. The style was a way of standing, or sitting, tilting one's head, and saying with a doubtful yet serious tone: "We don't know." "The data are not yet in." "The data are not good." "Could you design an experiment to test that?" "It hasn't happened yet." "How would you measure that?" "I don't know. I wouldn't do it." A person adopting the academic style dressed in a sport shirt or tweed jacket. He knew how to shrug his shoulders and smile in an offhand way, as if most things really did not matter. When he wanted to be tough, he spoke of getting his hands on some hard data. When he wanted to be gentle, he spoke of massaging the data.

I tried to avoid the academic style that I was increasingly exposed to. Possibly I was being rebellious; probably, the style simply did not fit. Two very smart students dropped out of our program at the end of the first year, and I thought it was because they did not like the style. They saw through it, or it did not allow them to be themselves, or somehow it was not worth it. The theory seminar met in one of the separate temporary buildings that was located far from everything else on the campus.

During the break in our three-hour class meeting, some of the students, and occasionally a faculty member, would play pool in an open lounge area nearby, where a pool table stood next to a soda machine. I did not know how to play pool and I often wished I did know, so that I might join them. Only later, when I ran into pool tables in lesbian bars and had a similar desire, did it occur to me that pool, like the academic style, was part of proving one was a man.

During the spring of my first year in the doctoral program, I gave up on communication policy planning. I told myself that planning was about politics: to be good at it, to get something done, you had to have the skills of a good lawyer. I wanted, instead, to be an artist and have freedom. I wanted to do work whose outcome did not depend on influencing other people, but depended more on my own internal decisions.

I was clearly uncomfortable with the scientism I found in graduate school: the concern with operational definitions, testing, hypotheses, and numbers.[1] However, I agreed with the basic philosophy. I, too, wanted to make statements that were descriptive of the world, if only in order to change it. Social science appealed to me because it offered the promise of finding regularities that might account for particular events of social life. I was interested in knowing what laws might apply: the law of oligarchy, the iron hand, the principle that frustration led to aggression. I did not want to hear that truth could become dated, that processes of history changed things, or that differences in culture made all knowledge relative. To me, statements were either true or not, and I wanted to know which. I was not used to thinking in terms of probabilities, or to viewing myself as part of a changing world, complicated beyond my imagination.

I am saying that I came to social science with a belief that social science provided a way to oppose superstition and illusion, and to find, instead, truth, which could be used to determine action. I was ready to swallow some fairly rigid truths if they would hold up. What I found were poses, scientific-sounding words, and sarcastic asides against those who might do work considered soft.

After two or three years of courses, most of the students in my graduate class wrote dissertations employing survey or experimental methodology. I wrote a qualitative sociological case study that was eight hundred pages long.[2] In the process of researching and writing it, I had to make up my own version of social science.[3] Much of my dissertation reflected values I had been taught, but the form of it was unusual: it read more like a fictional narrative than a traditional (conceptually organized) piece of social science. Partly for this reason, at the end of my final year in the doctoral program, after I had completed my dissertation, passed my orals, and finished all my formal degree requirements, a special meeting was called to review my case in my department.

As I drove toward the outlying building on the campus in which the meeting was to be held, I ran through a stop sign before turning into the parking lot out front. A uniformed woman police officer with shoulder-length blond hair and a gun in her holster pulled me to the side of the road and gave me a ticket. I felt shaken when I entered the building. Why was a lady cop carrying a gun on this peaceful campus? Why did she have to stop me? Why did I not see the stop sign? Only the answer to the last question seemed clear.

I sat down at the center table in the small room where four faculty members were gathering. I had previously been advised to do anything they said. The professor who had called the meeting sat at the head of the table and soon aired his complaint: I was not what the department was putting out, he said. They were a first-rate department known for a product. They trained survey and experimental researchers. The department had a national reputation to uphold. What would he, or any of the faculty, do if asked about me? I was not like them. I did not fit the type.

"If someone were to come here and write a book and submit it and say, 'Give me a Ph.D.,' why should we do it?" he asked, red in the face, looking straight at me. "You will not get a job," he then said.

"That's my business," I told him.

That summer, to satisfy this professor, I wrote an extra paper about the epistemological foundations of my work. Another faculty member who was kindly disposed toward me selected the topic as a way of bailing me out of our meeting. Two months later, he okayed my paper when I turned it in. I had done the extra requirement, although any reader of my paper could tell that I never really understood what my epistemological foundations were.

I was glad, in the end, to claim my degree, and I felt I had a right to it. Understandably, perhaps, I overlooked a few signs: the meeting, the paper, the importance of fitting in, the fact that I could feel threatened. Whatever anyone else thought, I was not worried about whether I would get a job and I did not see myself as different. If there was a standard product problem, I had yet to face it. Yet this problem would follow me, for better and worse, through teaching positions, research projects, and applications for jobs and grants. Repeatedly, I would think I was like everyone else, and that I was doing what was expected, and repeatedly I would be told I did not fit. It was not justifiable to be myself.

This book is about feeling more justified. It is about letting the self speak and about acknowledging individuality in social science. It is also about some of the problems that arise in that effort. My discussion draws on both my own experiences and the experiences of others, piecing these together, suggesting common themes, and ultimately offering an argument. That argument is for a more full acknowledgment of the self than is usually found in social science. It is for speaking personally about what we do.

2

Teachers

Approaches to Interpretation

SOCIAL SCIENCE is an interpretive activity: we make up stories to fit the world, or to make it intelligible to us. Initially, I did not have much self-consciousness about what went into the making of social scientific interpretations. Gradually, I began to learn. The most influential of my teachers were not professors I met formally in classes, but other people whom I came to know more closely during the time I was in graduate school. To speak of these people and what I learned from them is to speak of theory and method in social science. It is also a way to talk about how abstract concepts become part of a person's inner life.

Interpretation as a Gift

The first sociologist I knew well taught as a visitor while I was in graduate school in city planning. Timothy was British. His degree was from Cambridge in Moral Sciences. He could speak in a large lecture room about his studies of English widows or American poverty programs, but his style was most striking when he used his same interpretive skills sitting across from someone in a small living room or while talking when out on a walk.

Timothy would listen closely to a person speaking with him. Then, leaning forward, shaping the air with his hands, gesturing with his pipe, he would offer an interpretation back, an "argument" he called it. Timothy's interpretations were neither psychoanalytic nor broadly sociological, but they were a way of structuring thought that acknowledged social context and sought to identify regularities: "I would think what you feel is rather like mourning." "What if it's something like this . . ." "Do you think it is the insecurity of the attachment . . . ?" "I wonder if the new book requires a different sort of learning."

From listening to Timothy, I gathered that sociology ought to be given back to people it was about, who then would judge its usefulness. If an interpretation did not fit, the sociologist made up a new one. I also heard sociology as an oral tradition. Although Timothy wrote beautifully, his writing seemed to me an extention of his way of speaking. When I read his words, I heard him in them. In his writing and speaking, he dealt with similar themes, and in both, he spoke as if offering a way of thinking for his listener's inspection, offering it tentatively to see if it made sense.

Timothy's style was different from an approach in which a theory is presented in already tested form, as a statement whose validity the social scientist determines and then demonstrates. His approach differed from one in which theorists are viewed as separate from people theorized about, in which interpretations are assumed to be about others who are absent, rather than about those who are present and can comment on them. Timothy worked hard on his arguments: they had first to make sense to him; but they were neither remote, authoritarian, nor conclusive when he presented them to others. They were, for the moment, the best he could do.

At the time I first knew Timothy, I did not have much to compare with his style of analysis. Thus I did not know it was different from how others did social science. I felt his influence as if it stood alone. That influence was made all the more powerful by the fact that Timothy did not live at a remove from students, as professors often do. Because he was a visitor at the university for one or two quarters each year and needed company, he socialized informally and talked to people, reaching out with his words. In return for the favor of a dinner or the companionship of a walk, he would give back a sense of what he made of our lives.

I often felt uncomfortable with Timothy's interpretations because they were not exactly like mine. I worried that he was saying things too simply, or too much in a conventional style. I also felt that it was hard to respond critically to his views, since it seemed like I would be rejecting him in criticizing his interpretations, but that was the point. Timothy's statements were not apart from him, or from any of us, although I often wished they could be. I wanted explanations to stand alone and not be caught up in the confusions of social interaction. They should not reflect their makers or the contexts in which they were formed. Nonetheless, years later, when trying to structure my own thinking about social processes and research, I found that I wanted to be like Timothy more than anyone else. I wanted to give back interpretations that would be useful to people, and I found myself thinking about social meanings and social change in ways I had learned from Timothy, ways that emphasized the involvement of the interpreter.

Scientific Thinking

During the third year of my doctoral program, I began to see a therapist at the university medical school's psychiatric outpatient clinic. The clinic was located on the campus not far from where my department was housed. I went there twice a week and met with a woman who was well educated, sensitive, a psychiatric social worker, a member of the faculty of the medical school, and a university faculty wife. She told me at the start that she wanted me to learn that she and I were separate people: "I want you to know what is you and what is me." She also said that I thought I was invisible, that I spoke from inside of my feelings, and that I did not really know what went on inside other people. At the time I started to see Marcia, I wanted physical contact with people. She thought I ought to learn how to handle the middle: that space where people touch each other with the use of words.

I saw Marcia for four-and-a-half years. I am not sure if I learned what she wanted me to. At the time, I thought I was learning from her what I could not grasp in my classes. Marcia taught me how to think scientifically, or simply, how to think. She taught me how to assess evidence in light of theory, how to treat knowledge as incomplete, and how to view myself as worth knowing. Whereas in classes I shuddered at the word hypothesis, in therapy, although we did not use that term, we said things that functioned very much like it. We made cautious assertions about what might be true.

Marcia taught me how to put pieces together in a careful way. She taught me to view each thing I saw or felt as part of a larger picture that we could only know some aspects of. Most of the picture was hazy or not yet filled in. Event x was connected to event y in a particular way 70 percent of the time. The exact proportion mattered less than the fact that the statement was a guess, and the expectation that a guess, once made, landed in the context of a picture that was intricate and mostly beyond us: "Your friend is like a person who is starving and is handed some food and feeds it to the person next to her." "Something inside you is dead." "You wanted a reaction; that's why you threw the chair."

Theory lurked somewhere in the pictures Marcia taught me to imagine, but she did not often mention it. She did not say words like splitting, projection, transference, anxiety. She refused to call me a patient or a client. When the phone rang, she would pick it up and say, "I have someone with me." This was typical, I felt: the use of a common expression, even if it was awkward, in order to avoid a technical term that might categorize what I said or who I was. I felt that Marcia had long ago made a decision not to use standard terms or words from formal theories because she knew that the theories were not entirely right, and because she felt that the test of a good theory lay in whether it made sense in everyday language. Perhaps, too, she did not want to damn me with words that were not my own.

When I started to write my dissertation, not long after I began seeing Marcia, I decided to use theory as I felt she did. I decided to use more than one theory and to leave it embedded, or implicit, in my work. I would speak with regular words and make complex pictures that were not entirely filled in. When I began to teach classes a couple of years later, after finishing my degree, I thought about Marcia and felt, again, that I was acting like her: asking questions of the students, leaving things open-ended, depending on process.

Before I met Marcia, I had assumed that the purpose of

knowledge was to fit everything together in a clear-cut way. I thought I ought to know my own views very definitely, and that I should have an idea of the social structure of the whole world. Now I thought that the goal was to have a sense of the partial and particular nature of knowing. I thought that painting incomplete pictures was a good idea. I also recognized there was mystery involved. In viewing me as someone it would take time to comprehend, and then only partially, Marcia taught me to loosen up on the total systems way of thinking I had previously used.

In addition, because she interpreted me kindly and because she viewed me as not invisible, Marcia encouraged me to adopt similar attitudes in thinking about others. Thus social science became, for me, a style of thinking that was probabilistic, kind, open-minded, full of questions more than answers, and like having only a few pieces of a jigsaw puzzle. It also became something different from what I felt I was being taught over at the university.

In my courses, especially those concerned with method, social science seemed mechanistic and without values. When I looked at my therapy bills, I thought that psychotherapy was my investment in my graduate education. I had received financial aid for my university training, and what the courses did not teach me, I learned in therapy. Yet the system of private therapy supported the system of public education without acknowledgment. In my methods classes, clinical psychology was viewed as unscientific. These classes were taught by men, and men were in charge of the university. A woman did the shadow work of making my education human by making it relevant to me. I would later come to feel that this gender split was no accident.

A Fondness-and-Comfort Approach

I met Carl a year before I started seeing Marcia. After attending a few sessions of a class Carl taught on models in the social sciences, I wrote him a note and he responded, "Come on over. The water's fine." I never formally took a class from Carl, but I soon began visiting him once a month in his office. We started by discussing my dissertation. I had drafted a proposal. On it, Carl wrote, "You do not take yourself seriously enough," and, "You think doing a case study is easy. It's hard."

At the time I started meeting with Carl, I was also writing poetry and showing it to people instead of speaking with them. My poetry was personal and concrete and talked about my life. Carl seemed to like my poetry more than my academic language, which at the time was general, Marxist, and authoritative. When I finished my dissertation research and began to write my thesis, I wrote it like my poetry rather than like my previous academic prose. Carl seemed pleased. My dissertation was in the third person and it was about other people's lives, but it was concerned with personal-life details and was concrete and particular.

Carl used to say you could make poetry with ideas. When I first knew him, he was interested in the interplay of different kinds of theoretical models. He posed questions about how to choose among models: should you understand an event in terms of market dynamics, as a decision-making problem, a consequence of biography, or was it caused by chance? Marcia was teaching me to see things as pieces of a complex whole. Carl taught me to think about different formal ways the pieces might be connected. Given a set of data, one wished to imagine alternatives and to determine which type of model would best explain the data. There were different criteria for judging the fit of a model to the external world. However, and more important for me, Carl suggested that the choice of a model did not depend entirely on correspondence of data and theory.

Rather than asserting that one was choosing model x because of its superior power to explain the world, it might be more honest to admit that you were fond of it. Generally, people varied in their fondness for different types of theories. One person might like a Marxist view, another a psychoanalytic one, a third a symbolic interactionist vision. People were often just more comfortable with certain ways of seeing than with others.

If many theories potentially had use in illuminating a phenomenon under study, then why play a game in which any of these theories might lose? Instead of seeking to construct an interpretation that would be most correct, one might accept the various fondnesses argument and use the type of interpretation one liked best. The objective was to do so with the greatest ingenuity possible, or in a way that gave one pleasure or a sense of accomplishment.

Thus, theories were not necessarily competitive just because they were different. Instead of judging a theory, or interpretation, right or wrong, one attempted to understand how the theory worked and to see what one could do with it. One appreciated the agility of a good castle builder no matter what her persuasion, and one also had to decide which castles were to be one's own.

I felt greatly relieved on taking to heart Carl's fondness-and-comfort approach to the alternative theory business. That approach rested the choice of theories (the choice of how to structure vision) not on something external—on criteria attributed to an object of study—but rather on something internal: on preferences, struggles, characteristics, or likings of an investigator. In addition to a methodological concern, I shared with Carl a fondness for the view that people are not in control of what they do. This was a nonrationalist perspective, and it appealed to me because it was congruent with how I felt about myself. I saw myself as having many self-conceptions that had little to do with who I was, and I saw many plans of mine that were supposed to lead to clear outcomes (the idea of becoming a city planner, for instance) lead somewhere else. Having a view of my own actions that permitted me not to know everything in advance, like a fondness-and-comfort approach to intellectual choice, eased some of the pressures I felt to perform like a well-oiled machine.

Finally, I do not think I would have been influenced by Carl's ideas about social science if he had not viewed me personally with affection. I first saw Carl in a large lecture room, but I did not get to know him there. I got to know him in a small office where, again and again, we talked and I could change everything he said around to make it fit my experience. I could also be reassured by him that I, as a person, mattered most of all: more than anything I thought, or wrote, or did.

Whole Systems

My father was a writer. He woke early and wrote first thing in the morning. He viewed writing as a craft; one worked at it to be good. It was also a creative activity, like drawing or painting. My father taught me that there were principles to follow in writing that had to do with sentence length (short sentences were preferred) and with the use of words that an ordinary person could understand. He thought it was important to reach people by appealing to basic human emotions. Writing that worked in this way could increase understanding, help bridge gaps between different classes and cultures, and contribute to the cause of world peace.

Growing up in a household where a typewriter was often going at 5:30 A.M., I took for granted that writing was part of daily life. I also took for granted, and later adopted as my own, many of the ideological perspectives that informed my father's work.

In addition to believing in a common humanity, my father believed in the value of a labor movement. He read the paper each morning and positioned himself in relation to the politics of the world, much as others might position themselves in relation to the weather. My father interpreted almost all events in terms of their political import (did they improve conditions for the working man?) and in terms of economics (who had the wealth?). He also thought in a whole systems way. Using ideas about economics, politics, social forces, and occasionally personality types, he would explain almost anything. He taught his children to think like he did. I grew up feeling that my father's interpretations were not simply his; they were the way things were.

Not long after I enrolled in graduate school for my Ph.D., I visited my father back East. We got out of the car in front of my parents' house after having been out on errands. Although usually affectionate, my father, on this particular occasion, was mad at me for something. "Go back to your fancy professors," he said. The phrase hurt me. I was reminded that when I applied to graduate school for my doctorate, my mother wrote me, "We fear that you are becoming so esoteric that only a very few people will understand you. It will be like the Cabots who speak only to the Lodges and the Lodges speak only to God." My father had followed with a letter urging me to write my mother because she was crying at night and losing sleep.

This type of response was not unusual. In my family, people did not speak directly about their feelings. When we wanted to talk about ourselves, we talked about the world and gave opinions.

We used my father's political and economic type of analysis. Or we talked about other people (neighbors and relatives), using my mother's more psychological style. Both styles of speaking linked the term "I" with "think." We did not normally say "I feel" or speak of simple emotions such as sadness, hunger, or fear. Thus I grew up articulate about my opinions but without much of a sense of myself as a person apart from them. In later work, including my social science, I would need to get at the person with the feelings whom the rhetoric of my family overlooked.

My fancy professors were a problem for my father because they took me away from him and threatened to teach me other ideas than those I had grown up with. Further, in his view, universities stood for affectation. Members of my family were supposed to identify with common people and not put on airs.

When I left the East to go to planning school in California, my father, concerned with my safety, warned me about provocateurs. He had also done this when I was in undergraduate school and began going on marches. Provocateurs might be called Trotskyites, or they might be CIA agents. They got into the left wing and incited people to violence. After I was accepted into a doctoral program, my father warned me about academics: these people had the worst kind of meetings. Their meetings took longer than anybody else's, and they didn't come out of them with their problems resolved. College professors, my father felt, had no sense of the reality of the politics they were engaged in.

I think my reasons for feeling uncomfortable in academic meetings are mine more than my father's. However, in other areas I am less clear. Because my father taught me many things that are still important to me, and because he died soon after I finished graduate school, I often feel I am my father rather than myself. There are, indeed, some traditions of his that I carry on. While I do not write of politics and the labor movement as my father did, I have my own version of politically virtuous work. However, my approach requires personalizing where my father generalized. If I do not make things personal, I become lost. Repeatedly, I must construct a sense of myself in order to feel different from others and capable of living. A basic reliable sense of self is not something I can take for granted. I think my father did not have that problem. Thus, issues of self were not central in his work, while they are in mine.

I wake up in the morning later than my father did, but I write first thing and I do it in a similar persevering way, seeking to make statements that are faithful depictions of the world. In getting a Ph.D., I thought I was taking a route to a stable, salaried job so that I would not have to search for new work all the time like my father did. As it turned out, I have ended up searching. I seem to have inherited my father's idealism, even if my particular commitments are different from his. Idealism often ignores a concern with economic security. My father wrote history and biography. He had no use for social science. However, he did not oppose my doing it. When I picked my dissertation topic, I consulted with my father, seeking his approval. I thought my topic was my own. He approved on the basis that my research would be in the interests of the labor movement.

For other kinds of approval and other ideas about how the world might be understood, and how I might think about myself in it, I needed other teachers. They came later, after most of my father's work with me was done. To a large extent, themes from my father's early teachings carried through and affected whom I later chose to learn from. My subsequent teachers had to be good in some ways my father was: they had to be kind, and concerned with ordinary life, and they had to value intelligence. They also had to be different from my father, for I needed to learn about a world that was less ideologically structured than his, and I needed help in feeling that I was of value in my own right. The teachers I chose influenced the nature of the social science I learned to do. Each displayed a unique relationship between self and work. Seeing that relationship was important for me. It was vital that I realize that social science met personal needs and that abstract thinking was not only about the world, but also about the way different individuals related to it.

II

Forms of Expression

3

The Vulnerability
of a Writer

IN THE PAST DECADE, social scientists have become increasingly reflective about the way rhetorical devices structure knowledge.[1] We are advised to look at the forms we use to describe events: the story, the drama, the scientific study. These forms are viewed as more than techniques for presenting findings, for they affect what we know. At the same time, there has been a renewed interest in the subjective basis of social scientific inquiries.[2] The challenge is to understand how our personal experiences affect the tales we tell. Some of us have become increasingly dissatisfied with the tone of remote authority commonly used in the writing of social science[3] and with the way the personality of an author gets lost in social science texts. There is something about the social scientist's self that we seek to express, and to hear, more fully in our works.

Traditionally, as I have noted, social science has viewed the self of the social scientist as a contaminant. The self—the unique inner life of an observer—is a variable we are taught to minimize in our studies, to counter, to balance, or to neutralize. We are advised to avoid self-indulgence, not to view other people in our own image, and to speak in a manner that suggests that what we know is not particular to our individual way of knowing it.

I wish to suggest that the self is not a contaminant, but rather that it is key to what we know, and that methodological discussions might fruitfully be revised to acknowledge the involvement of the self in a positive manner. The self is not something that can be disengaged from knowledge or from research processes. Rather, we need to understand the nature of our participation in what we know. The problem we need worry about is not the effect of an observer's inner self on evidence from the outside world, but the ways that the traditional dismissal of the self may hinder the development of each individual's unique perspective. The following discussion focuses on the struggle to articulate a personal point of view in social science and on the experience of vulnerability that often accompanies that effort. In it, I use examples from my own experience to illustrate more general points.

Inner Voices

I write both social science and autobiographical fiction. For some time, I have noted a difference between the two. In each form, I attempt to grasp experience. In each, I deal with both my own experience and that of others. However, in my social science, I find I am much less sure of what I want to say than in my autobiographical writing. I lapse more often into making polemical statements and later feel I must cross them out. I have a harder time being direct. I find it easier to write in the third person, and to speak of others, than to write in the first person and speak of myself. In my autobiographical accounts, on the other hand, I find I focus far more easily on details that will be revealing. I feel less confused than in my social science. I may not know what I want to say in advance, but I am more confident about my right to say it.

Recently, I have taken to writing social science in the first person in order to feel more present in my work and less lost. However, I notice that I pull back each time I speak of my own experience. I make a brief statement beginning with "I," then follow it with a "we" or a "one." I feel grand emotion when I write of "we" or "one," as if I am speaking for the world, but when I reread what I have written, I do not find it compelling. My "we" seems to get caught up too frequently in old battles about how social science ought to be done. Several months ago, I tossed out 80 percent of a manuscript that was a rough draft for this book. I wanted to omit pages where my "we's" and "one's" overtook my "I." I could see how often I kept running from myself. I excluded discussions of the ways universities value smartness and why these are discriminatory, for instance; discussions of power games in the development of new theories; explanations of how a muse functions, generally speaking, for an author; lists of differences between the disciplines in how they sought to correct for the self.

I felt I was not present in my pages of larger-scale and more institutional rhetoric. I was not in these pages enough, or not in them directly and, thus, not brave enough in what I said. When I reread my generalizing statements, I saw that I could be clever and sarcastic. I could make arguments that sounded reasonable and arrogant. But these were not things I valued. They were not worth fixing up or showing to other people. I was mindful of the fact that it had mattered a great deal to me that people had liked an article I wrote several years ago titled "Beyond 'Subjectivity,' " about writing *The Mirror Dance* (see chapter 13). That article meant more to me than the previously published book because it spoke of the process behind the book's third-person voice and it made me less hidden than I was in the formal study. When writing my later book manuscript, I was afraid I would not do as well again. I would not be able to be as candid.

Why was I so afraid in the later study? I was used to writing in the first person in my autobiographical work, and I was now arguing for acknowledgment of the self in social science. Why, then, could I not speak of myself without running from my own statements? Why did I have such a hard time being direct? Who was I fooling? Who was I concerned with letting down? Why did I think it was others rather than myself? Why did I always think it a lie when people said they wrote for themselves? If I were only writing for myself, I would not write. Who did I write for? Why did I insist on writing for people who did not want to hear me? Why did I find it so hard to speak?

These are questions from my internal voice. My inner voice is full of questions, fears, and negative judgments about myself. My outer voice, on the other hand, speaks in statements. It makes cautious assertions that hide my inner uncertainties. The effort it takes to fashion the outer voice tells me that the central issue for me in both my autobiographical and social science writing is the issue of hiding. In both forms, I seek out the perfect word or the correct idea, trying to make a smooth surface that will not betray me, that will not reveal who I am: a person who may not be acceptable. At the same time, I wish to be revealed; I want others to know me. But I want them to know me in a specific way. I want them not to judge me and find me lacking.

I think it is possible that people in their writing are especially vulnerable to external judgments, or to judgments felt to be external in their source. Academics have particular ways of hiding in their writing in order to avoid such judgments, especially when the judgments may be negative, and all these ways have consequences for the individual's sense of self. As social scientists, we use standard terms and depersonalized voices that camouflage the self and make it conventional in order to make it acceptable and in order to communicate our thoughts. Yet curiously, these very attempts at protection exacerbate vulnerability problems. For if a work that was supposed to protect the self is criticized by an outsider, the greatest threat is not that the work will be found faulty, but that the self behind the work will be exposed. The original sin—the individual who was so dutifully covered up with proper language, theory, and good form—will then no longer be safe from harm.

I think that in social science we write to protect as much as we write to express the self and to describe the world. Although we speak of protecting others—usually, the people our studies are about—the main object of our protective strategies is always our selves. I may wish to protect a sense of myself as a moral person, for instance: a person who does not do others harm. Or I may wish to protect a sense of myself as an intellectual person: a person committed to figuring things out. To speak of desires for protection and acceptance is different than to speak of power as a motive in social science. It is different than seeing social science as a quest for dominance—a contest to determine whose ideas, or whose person, will prevail. Acceptance, as a theme, may emerge more from female experience than from male experience. Similarly, a concern with vulnerability may be more commonly a female cultural concern. These concerns are also specifically my own.

I think we have less awareness of the negative consequences

of our self-protective strategies in social scientific research and writing than of the positive consequences. When we write in distant and impersonal ways, we underestimate the extent to which a distanced stance can alienate us from ourselves and each other, and make us less able to speak the truth of our experiences as we move farther from them. The intellectual apparatus we set between ourselves and what we know (the categories, theories, and concepts we use as if these were unrelated to our needs) can also alienate us from the experiences we seek to describe, as can the negative attitudes toward the self that cause us to bury our feelings. Strategies that emphasize conformity in our views further encourage alienation and distance from experience. By writing to fit in, or to blend, with what has been done in a field or a discipline, we contribute to a general climate of fear concerning what might happen were our individual subjectivities to be given more room. What would happen were the world truly to be seen according to multiple and different points of view?

The fear usually spoken of is what would happen to social science were differences in individual perception stressed. How would people be able to build on each other's knowledge? How would anarchy be avoided? The implicit fear is of what would happen to each one of us. Will we be rejected because we are different? Will there be anything to belong to? How will we be able to take steps to safety if there is no agreement about what is true?

I do not fear anarchy as a result of the expression of individual perspectives. I think social life and social science both demand a great deal of conformity. Even when we try to be different, our individual expressions look a lot like one another. I am concerned, instead, with the difficulty of developing individual views and with the issue of vulnerability. Personal expressions in social science often leave one feeling vulnerable. To avoid vulnerability, one often seeks to protect the self by limiting the extent to which one discloses personal information. I would like to be less concerned about being vulnerable in social science than I have been. I would like to write social science more like autobiography and not worry about lines between the two. My autobiographical writing feels better to me than my social science because I am less hidden in it and because it is more intimate and closer to my experience. However, I have difficulties treating my social science as something I can be the focus of. It is not only for others that social science looms behind a gate posted with a large sign that reads "Leave personal items behind. Make sure that everything you say is true. Remember that the truth of your individual experience is not enough."

Given that such a sign marks entry, which means that it is part of one's socialization, efforts to view the self as enough—as an experience worth describing and as a legitimate source of knowledge—face great difficulty. My efforts to speak of myself not because I am generally applicable (like a theory) or because I am an instance of a more widespread phenomenon (like a piece of data), but because I am someone in particular, are difficult efforts. They are fraught with all the prohibitions of a field built on denying the self. No wonder that ideas for how to speak of the self must come very often from outside of social science—from the arts, autobiography, and ordinary life. The goal is to use a variety of approaches to articulate the self within social science. However, the process must often be more far-reaching, for, within social science, our ideas about the self serve largely as guides to self-censorship. They tell us when to view the self as irrelevant, biasing, or misleading, and when to separate the self from what is known. What we need are examples of how to use the self more fully in order to develop our capacities for insight.

Gender, Writers, Inner Spaces

My friend Helen grew up with a mother who was a writer. Her mother worked behind a closed door, locking herself in her room so that her concentration would not be interrupted. My father wrote with the door to his study ajar. His open-door habit came from years of writing in the city room of a newspaper and in the headquarters of a union. When he wrote at home, being interrupted by a child, or by a doorbell, was not jarring for him. I often felt that my father welcomed household interruptions. They did not last long. They were a relief from the pounding out of words he normally engaged in. They were an opportunity for company.

I think my father's attitude toward writing was affected by

the fact that he was a man. Thus the children interrupting his work at home were not in the same relationship to him as were the children who interrupted my friend Helen's mother at work. Because he was a man, my father did not have to fight for room for his writing. He did not have to worry about whether doing his work meant he was holding out on his children, or to consider it competitive with other obligations he ought to fulfill. Writing was his central obligation. It was the way he earned his living. It threatened to overwhelm other things, not the other way around. Helen's mother, I suspect, felt that her children and household might overwhelm her.

Helen learned early writing habits from her mother. When I first met her, she felt a need to close a door while writing and to moan from the other side of it when her work did not seem to be going well. When I first met Helen's mother, I found out even more about differences between Helen's and my formative experiences. Helen's mother lived in a world where to be a writer meant one subscribed to *Writer's Market,* took courses on writing and literature, kept up with who was famous in contemporary literary circles, admired Ernest Hemingway, sent off short stories to magazines, and generally fancied oneself a writer. This was very different from my father's approach, in which writing was viewed as disciplined factory work and as the equivalent of digging a ditch. As children, we were repeatedly told that a man working in his study writing burned up as many calories as a man digging a ditch. The calorie story was a way of reminding us that my father did real work and was not unlike any other working person.

For Helen's mother, writing had to do with artists and bohemians. Her sense of who would carry out the revolution (an artistic avant-garde) was as different from my father's (the proletariat) as was her economic situation: that of a woman raising children without a wife. As I heard her speak one evening while sitting in Helen's living room, I could not help but feel that Helen's mother's idea of being a writer was light and airy compared with my father's. It reflected the trappings of being a writer without the bread and butter of it.

Before I left that evening, Helen's mother, as if sensing some dissonance between us, asked whether I had taken any courses in writing novels, since I was then working on one. Her implication was that I had no right to be writing a novel if I had not first learned the proper way to do it. The issue of "the proper way" also arose in my relationship with Helen. At the start of our friendship, I gave her several autobiographical stories I had written and a draft of the first chapter of *The Mirror Dance*. She was horrified that there were at least half a dozen grammatical mistakes on each page. I was shocked that such a thing might matter. My father had not taught me about grammar, but rather about extra words (remove them), the common man (be for him), and Hemingway (be sure never to read Hemingway or you might end up writing like him). My father told me periodically that he never read Hemingway because he felt that in his generation, writing like Hemingway was a contagion: everybody did it. If you read Hemingway, it was almost unavoidable that his style would affect yours. Since I tended to see my father as unafraid, his fear of Hemingway struck me as odd.

In retrospect, I think my father had his own version of Helen's mother's closed door. It simply was not a door to a physical room in a house. My father, like Helen's mother, was concerned with selectively letting in and warding off stimuli from the outside world. If he was not concerned with grammar and literary circles, he was, no less than Helen's mother, committed to imposing his own approaches to doing work and using words. These included his workaday attitude toward writing, his sense that words should disappear, leaving only the meanings they conveyed, and his commitment to the idea that the social facts words stood for mattered most of all, and that you could not learn those facts in an English class. With his various prejudices and habits, my father created an inner world that was vital to sustaining his work, much as Helen's mother created a world with her closed door and her sense of literature.

In thinking of these two different writers, I am struck with how important the creation of an inner world is. In social science, I think, we do not learn very much about how to build from within. We learn how to conform. We are encouraged to speak in generally acceptable styles, rather than to speak in ways that are our own. The ability to speak from within takes nurturance. It requires the use of one's own words rather than the use of currently fashionable words in one's discipline (the deconstructionist vocabulary now popular in many of the social sciences, for instance). It also requires actions to protect the inner space in which one's ideas can take form.

Inner spaces need protection both so that inner development can occur and because of rejections: ideas sent back, articles, books, stories, applications that are returned by a world that would rather not hear one's particular point of view. My father, Helen's mother, and experiences of my own remind me that rejections come often for a writer and that they hurt. Helen's mother repeatedly said she thought she was supposed to get used to the rejection slips that came in the mail, but she never did. I saw my father's pattern of constantly sending out his materials and getting little response back, and I felt all that self-promotion required a toughness I did not have. However, academic as well as literary and political writers must send things out; they must offer their ideas and inventions to the world and then deal with the response, including the response of rejection and that of criticism aimed at shaping one's work into a common mold.

I have always found rejections, and the criticisms implicit in them, very hard to deal with, a difficulty that has increased as time goes on. I think rejections are especially troublesome for people whose work and self are not separate. I know I feel penetrated by each rejection I receive. Each time, I have to build up something inside myself in order to continue. Usually, the thing that needs rebuilding is some idea about who I am and why there might be value in what I have to say. Rejections have the effect of wiping out my sense of value and causing me to wonder why I ever was so presumptuous as to send something out in the first place. They cause me to feel humiliated for exposing my expectation that I should be accepted. Because I so often do inner rebuilding (convince myself, again and again, that it is worth writing the next book or applying for the next job), I am surprised, each time, that the outside world can get to me as much as it can. I am surprised by how unprotected I am. I tell myself that I receive rejections, in part, because my work does not fit a standard type, but this does not help me very much. I suspect I am not unusual in my difficulties with rejections. People generally need buffering from a larger environment when it is hostile or disregarding.

However, in social science, we do not normally talk about

our needs for protection against an invasive or indifferent world. We do not talk about the fantasies we have about ourselves that help us to keep going, or about the ways we can build social ties that will sustain us in work we can feel is our own. We speak, instead, about which ideas we should use, or which methods, and we speak of these as if they were detached from ourselves. This tendency to overlook the self can lead us to take for granted the strength that is needed to continue to offer pieces of ourselves, through our work, when no one else seems to care. Only when we stop detaching self from work will we be able to acknowledge such strength and to foster it, and will we be able to work against the surface conformity that stifles many of our ideas. This conformity, which, to a large extent, is a product of desires for collective protection, causes us to make our work look like that of others. It leaves us with a less rich sense of experience than we might be able to obtain were our individual perspectives discussed and valued. It also encourages a fear of who we are as individuals, and of who we might be, were we to speak in more self-connected ways. Speaking from the self, as exposed as it may feel, seems to me a very worthwhile endeavor. Although rejections will come, the quality of what is offered outward may slowly change when the self is acknowledged more. In the process, the criteria for judging what is legitimate in social science may also change.

Personal Sources

Five years ago, a foundation that funds work in social science invested two years worth of money in me. I had proposed to write a book-length manuscript titled "Fiction and Social Science: A Methodological Inquiry" that would further develop ideas presented in *The Mirror Dance.* In an essay at the end of *The Mirror Dance,* I argued that fiction and social science had a great deal in common. Now, funded by the foundation, I put finishing touches on a novel I was writing and began a 250-page academic manuscript. I submitted the academic manuscript, at the end of the two years, to the foundation and to several university presses. Reviewers' comments for the presses suggested that I revise the work and offered recommendations on how to do so. The substance of their critical comments fell into two areas. One was the area of definition: the reviewers tended to have their own ideas on how to define similarities and differences between fiction and social science, and they held their views with conviction. I felt that the reviewers each had their own book on the subject of fiction and social science in the back of their minds, and it therefore seemed not to matter very much that I was offering mine.

However, what was most striking to me was the fact that the reviewers of my manuscript reacted strongly, and often negatively, to the structure of my work, a structure that relied on my own research and writing as the primary source for my methodological statements. I soon came to feel that the reviewers' structural criticism was more fundamental than their definitional criticism and that was perhaps what gave the more objecting readers license to discount so passionately much of what I said. These reviewers were bothered by the fact that except in a brief introduction to my manuscript, I did not refer to studies or novels written by others, as is usual in academic work, or to studies about sociologists or novelists. Rather, I spoke as a person who had read and practiced in the area and who was trying to make sense just as herself. I spoke, in other words, based on personal authority, using my own experience as a guide. In retrospect, I think I may not have done this terribly well (I tended to use myself and hide myself at the same time), but the general message that I was my own authority was clear.

I had offers from several publishers that if I could revise my "Fiction and Social Science" manuscript along conventional lines, the work would be publishable. I wanted to be published. However, after sitting with my "Fiction and Social Science" manuscript for a year and a half, I came to the conclusion that I would not revise and resubmit it. I did not want to change it along the lines the reviewers proposed. I concluded that I could not add footnotes, or external references, without changing the structure of authority of my text from a structure built on the specificity of a personal source to one that drew on more general sources. The use of external sources seemed false to me since these were not truly the basis of my thinking. More importantly, the personal parts of my manuscript were the sections I liked best and I did not want to lose them or their logic. These sections grounded the whole manuscript emotionally for me: they grounded it in experiences I could feel.

At the time I wrote my manuscript, I especially needed emotional grounding. I had begun this manuscript soon after a therapist I was seeing was diagnosed with terminal cancer. While I was working on the manuscript, my younger brother died. In part because of these events, I wrote my "Fiction and Social Science" manuscript on automatic, stringing words together, hoping they made sense, following suggestions made by friends about how to complete my sentences. Because I felt generally alienated from what I was writing, this manuscript was harder for me to revise than any other piece of writing has ever been. As a result, I did not revise it much. I was therefore surprised that the manuscript was readable in the end, and more surprised by the reviewers' criticisms of my use of personal authority, since that was the one thing that made sense to me. Some of the reviewers even suggested omitting the personal portions of my manuscript entirely. I felt my critics did not understand my work. They did not see how it functioned structurally. However, in the long run, I realized there were things I also did not understand. While I was writing my manuscript, I had thought I was writing about fiction and social science, but I had, more basically, been writing about social science and the self. My style—the "how" of my work—was more important than the subject of it.

Subsequently, nearly two years after completing "Fiction and Social Science," I decided to write another book that would take on the self issue more directly. I wanted to do in spades what I had done before and been most criticized for, but, this time, to hit my readers on the head with it so that my use of personal authority would not be seen as a weakness of my volume, but as its strength. The new book would be titled "Social Science and the Self" and would be personal without apology. Why shelve the first manuscript (possibly half good, already half developed) and start another? I think I take rejections hard, even when they are formally "revise and resubmits." I also tend to feel that I cannot rearrange something I have written into another shape or form, or organize it according to other principles. Thus, if something different, or better, is required, I start again. I also think that, in this case, I was influenced by the deaths of my therapist and my brother more than I knew at the time. Their deaths had made it hard for me to write my manuscript, since writing, for me, requires feeling and I did not want to feel at that time. Two years later, I looked at my new title, "Social Science and the Self," to which had been added a subtitle, "Personal Essays on an Art Form," and I thought, "Well, how am I going to pull this off? How am I going to avoid getting alienated from this one?" I thought about many things and, among them, found it useful to think of my brother.

Possibly I simply missed him. Possibly I needed a sense that someone else was telling me what to do. My brother reminded me of inner struggles people often have and of the need to talk about them, directly and in writing. My brother had been someone who used to turn to me for help, but who did not turn to me that last time when he died. I wanted to write a book that my brother, had he been alive, would have said after reading a few pages, "Yeah, I see what you're doing," meaning, "Yeah, I see why it's important. Keep it up."

I would like to think that the death of my brother and of my therapist shocked me to my senses and made me want to do a work that would deal unapologetically with something central to me: the relation between self and work in social science, the relation between myself and my work. Perhaps nobody had to die for me to focus on this topic, but it is sometimes hard to justify writing about issues that matter to you, in the absence of someone else in whose life such writing might also matter. It is easier for me to think of my brother as worth saving, and worth talking about, than it is for me to think of myself in those terms. So I sometimes said to myself, and explained to others, "I am writing this book because in a more personal world, a world in which people do not have to hide, my brother might not have been hit by a train. He would not have been walking alone on the tracks that rainy morning, feeling he had no alternative but to end his life." In my mind, I would see him there still, looking up helplessly before he died, looking in my direction, waiting for something to change, waiting for me to change it.

My brother and I sometimes read each other's personal writing. Once I loaned him a book of my poetry. He gave it to a close male friend of his and I was shocked, although I felt it meant he was proud of his sister. He gave me a copy of a journal he kept and I loaned it to my therapist. I think we each wanted to be understood as related to the other, even if we did not always get along. I remember my brother once wrote a letter to me in which he commented on some stories I had sent him and on our father's preference for short sentences: "I see you are writing long sentences and I think that is good. They let you breathe more and I hope you will keep on doing it." In the same letter, he discussed getting a new pair of shoes and being self-conscious about the height of the heels when he walked in them out on the street. He thought everybody was looking at his shoes and laughing at him. The thought was painful. His conclusion was an observation: "One's life can be on the line for all practical purposes in getting dressed." I like that line because I also feel that way and because it sounds like my brother speaking.

I will never know another person more like myself than my brother was. He could write far better than I can, but he chose not to write for a living or for public consumption. In my basement, I have four boxes of photographs he took and notebooks he kept. I keep thinking that I will find my brother in the pictures and notebooks he left behind. I keep thinking that the way I feel about my brother's death should influence my social science. It should make me more attuned to the importance of my feelings, for I think that my brother died for lack of emotional understanding and because he was trying to live up to external standards of independence and success. My brother, I feel, would want me to learn from my feelings and to measure my worth from inside. He would want me to go against the destructive forces that can sometimes tear one down, even from within. In an important sense, my brother's death helped me take the risk of writing this book in a personal way, despite my feelings of vulnerability.

4

The Presence
of the Self

THE EXPRESSION OF AN INDIVIDUAL PERSPECTIVE in social science
is a difficult accomplishment in part because individuality is theoretically unpopular. The social sciences tend not only to view the
self of a researcher as a contaminant, but they also view the selves of
people studied as invisible. In sociology, for instance, the self is often construed as a hollow core, best understood as a reflection of external forces that act upon it.[1] In experimental psychology, the self
is something we typically know only in terms of measurable external behaviors, or we know it in terms of cognitive processes that do
not comprehend the whole self. In economics, the self is expressed
through preference functions: choices about what is valued over
what. In political science, the self is something symbolized in
rights, powers, and acts of political participation. In depth-
analytic psychology, ideas of the self become internal, but they also
become extremely complex: the self appears as a structure (id, ego,
superego), as a pattern of defenses, or as a pattern of internalized object relations. In the field of self psychology, where the self is the
subject, it is still difficult to grasp a sense of the self as a unique inner experience because the concern is with demarcating processes
all selves share.[2] In anthropology, we are reminded that the self
takes different forms in different cultures. Recent anthropological
emphases urge us to see the self as fragmented and multiple, reflecting the different roles people play in their various relationships

with community life.[3] In history, we are encouraged to view the self as reflecting historical circumstances and as changing over time according to prevailing definitions, rather than as an experience whose meaning is constant or highly individualized. Because the social sciences are generalizing sciences, there is a natural tendency to deemphasize the particular and internal nature of the self and to see the self in intellectual terms. This results in descriptions of the self that have little to do with everyday experiences of individuality, in contemporary American society or elsewhere.[4]

I do not wish to argue here for a view of the self as fixed (unchanging) or nuclear (rigidly structured around a core). Nor do I think that the self of an individual is separable from social life—from the web of a person's relationships. Definitions of self are always culturally, and subculturally, relative. However, in the context of social science, I want to argue for a strong definition of the self as an individual and inner experience. I want to speak for a view of the self as a central organizing mechanism within each individual that is experienced, to some extent, as unique. I emphasize a view of the self that acknowledges inner experiences of individuality both because I think these experiences are important sources of knowledge, sources traditionally minimized in social science, and because I think a sense of individual uniqueness is often hard won. Such a sense is frequently a difficult achievement that is felt as precarious by the individual and that is experienced as a struggle: a struggle against being like everyone else, a struggle to hold together or hold up, or a struggle simply to feel that one has a self. Often other people appear as more clearly "themselves" than one is oneself because others are seen from a distance, whereas the self and its confusions are known close up and from within. Enmeshed in internal conflicts and associated with a sense of inner invisibility, one's own self often feels less defined and more difficult to grasp than that of others, or than the shape of any other external reality. The difficulty of grasping inner reality may be one reason for our common external focus: the outside world is easier for us to know and so we focus more upon it.

Although struggles for self may be difficult, and even unselfconscious, and although these efforts may vary in different cultures, over time, and between and within individuals, I think we can still talk about something called the self. Indeed, I think we must, not because the self can be known once and for all, or because it ought to be known in any specific way (there are many theories about inner selves),[5] but because discussions about the self can be useful in grounding our understandings. Further, if we do not talk about ourselves, and if we acknowledge only the general forces that affect who we are and what we know, we ignore the full reality that informs our work. An important part of that reality is internal processes of self-building.

There are many kinds of self-constructing activities, and these produce very different selves for different people—selves that look different on the outside and feel different on the inside. Gender socialization is one important factor affecting the selves we have: male selves, to a significant extent, look and feel different from female ones. The male self in social science is, I think, largely what we know; it is possibly a more straightforward construction than the female self might be were it more fully expressed in our studies. The male self is more straightforward (more simple) because men are socialized in our culture to take for granted a great deal about their underpinnings that women cannot take for granted, since women often are the underpinnings. Women learn to lose themselves to others more often than men do. They are familiar with the work of keeping others afloat, and especially familiar with the work of caring for the weak, or wounded, egos of men.[6] However, men may learn how to do emotional support work, too, when their sense of self fails, or when they are called upon to help themselves, or to help others, in ways that have been traditionally viewed as female—ways that respond to the internal emotional needs of the self.

Often only when the self, or the sense of self, becomes precarious do we learn that having nobody home is not a good idea, that theories denying the importance of an inner self do not serve us well. Having one's internal organization reduced to a minimum, or to something felt as extremely tenuous, absent, or fragmented, or to something felt as not specifically one's own, is frequently a painful experience, one to be avoided and recovered from.[7] It is not an experience to be celebrated as essential to the self, as some contemporary accounts suggest.[8] Recovery or rebuilding activities are ways the self survives in spite of threats to its organization, and such threats may occur frequently because the self is not alone, and because often the major forces affecting the self have little to do with the self's own integrity. The doing of social science can be a way to further organize the internal life. It is not the only way, and it does more than achieve internal order (social science also organizes the external world), but it very much reflects internal imperatives. Needs of the self for various kinds of coherence, intelligibility, and response continually affect our interpretations of others and of the external worlds that our studies describe. That the self has needs, and that social science reflects this, is something we do not normally discuss at length, but I think it is an important topic. One need of the self is for self-expression, for articulating an inner sense of individual experience.[9]

I was not raised to think in this way about the self. I was raised familially, and socialized professionally, to overlook the self, especially my own self, to take for granted that I would hold up, and to view people generally as less important than their social circumstances. At the same time, I grew up looking to individuals who seemed to be rooted within themselves, and I took special pleasure in learning from and knowing them. These were people who did not fade away or become some version of the sociologist's hollow core, the psychologist's set of reportable behaviors, or the anthropologist's idea of a person inseparable from a specific culture. I continue to look for people's selves in their writing, and in their social science writing, in particular, where the self is often especially hard to find. Perhaps I want people's autobiographies a bit much. I know I want a sense of the person behind a work, and I will seek out such a sense even if I have to make up my own version of an author as I go along. I think it would be better if I could be told more about the author from her point of view, for then my invention will probably be less off the mark.

Looking for a sense of an individual person in a work is one way to have a sense of security in relation to others, and it is one way to comprehend some of the sources of social science. However, it is a way that is often pushed aside and dealt with very minimally, in part because the self is internal and enmeshed in feeling, and social science has traditionally been about an observable external world that can be conceptually known. Ideas about the individuality of the self are worth paying attention to, I think, not because the individual is, in reality, a separate being who stops or starts at her skin, but because the individual has inner experiences. These experiences may not be felt as entirely one's own, but they will reflect one's particularity. People are differently situated in relation to their environments and, as a result, have different possibilities for experiencing the world and for knowing themselves.[10] The self is not entirely individual or internal, but it is this aspect that draws my interest.

Traditional notions about the self in social science, and some modern ones, seek to ignore individual differences and the peculiarities of inner experience, especially when the subject is the social scientific observer. Such an observer is advised to overcome individual differences, to interpret the world in fairly standard terms, and to counter distorting effects of the self. I wish to suggest that the inner experience of an observer, however peculiar, can also be put to good use in social science, for it may be a source of fruitful ideas and illuminating perspectives.[11]

In the social sciences, we are used to claiming that the experiences we describe represent external realities rather than realities of the self. To map those realities, we use abstract concepts that are organized according to a formal sense of balance and logic. My discussion here, although it has elements of the abstract, is a different kind of mapping. It uses concrete experiences to call attention to certain aspects of reality. I do not seek to identify a middle ground in describing the self or the issues of social science, but rather to grasp for a perspective that is mine.

This grasping for my own view creates a picture that is, by many standards, not quite right. However, that picture may provide a lens that is helpful for envisaging other realities as well as my own and that clarifies methodological issues. It is in the nature of the self, I think, to work in this way—not to map the world in a formally correct sense, but in a manner that highlights some things more than others and that connects events in particular ways. The presence of the self is organizing of reality and of how we see events. The self, in coalescing perception, produces a picture that has an order determined, significantly, by emotional emphases. Strategies of using individuality, subjective vision, or a more full sense of the self in social science may cause us to produce different types of compositions for understanding our experiences than more standard strategies do. They may result in different forms of expression, but these forms will have their own coherence and themes, and they will have a structure that is truthful to the degree that it faithfully represents the experience of an observer. That experience includes the observer's inner life: the dreams, the imagery, the sense of "how it is for me" to be here, to see this, to relate it to you.

Inner experience is no less real than observation of the outer world, but it does require more acceptance of what is deeply felt, and it requires tolerance for images that do not fit standard notions. It is important, I think, to become less silent than we have been about our inner experiences in social life, particularly when these experiences seem wrong, or deviant. It is important to be present in our studies and to create forms in which we can be known as specific authors. One consequence of being more specific about the self is that, in the end, one becomes more general. One person's idiosyncratic experience speaks to the experience of another. People find likeness despite difference, and they find it all the more when more is said about the self. To encourage a social science based on individually different expressions is not, I think, ultimately destructive to the building of joint visions. It is simply a different way to go about developing such visions, and it is potentially a more rich way than others that emphasize likeness, or a common view, or that leave the definition of the self unclear. Not all social science need be written in the first person or with a strong sense of authorial presence, but I think that some of it certainly would be more interesting if it were written in this manner. Articulation of the self is something to be developed, rather than avoided, in our work.

5

Speaking of Writing

OCCASIONALLY, friends ask me to speak to their classes about writing social science. I was surprised, at first, to be called in for the specifically textual part of a course. Writing, I felt, was an activity to be engaged in but not spoken of. It was essentially a private matter.[1] Further, I felt that to speak of writing meant I would have to discuss forms and devices like a literary critic, which was not something I valued doing or knew how to do.

I felt I had to develop my own approach when I went to the classes. One of the first invitations to speak was in a course on feminist ethnography. The woman teaching the course told me beforehand that she used *The Mirror Dance* in her syllabus as an example of an objective account, as opposed to a subjective one. I felt taken aback by her comment. I had not before realized how much I wanted *The Mirror Dance* to be seen as subjective. I also thought I probably should have written the study more personally to make its subjective nature clearer. I felt a desire to rebel at the categories subjective/objective, but if I had to be placed, I wanted no part of the uninvolved objective position.

When I met with the class, I was determined to emerge from behind my text. Thus my discussion of writing was personal. It concerned my problem of not being able to write *The Mirror Dance* for two years, a discussion later elaborated upon in more detail in the article "Beyond 'Subjectivity.' "[2] In both the class discussion and the later article, I was glad to present a personal story. The strength of that story lay in its discussion of

49

specific details of my experience. Nonetheless, when I presented my experience, I subsumed the particular details under the rubric of a broader story line. That broader story was one of success. In both the class and the article, I spoke of saving my study from becoming lost. "Here is how I made it all work out," I said. "Here is how insight about oneself can inform one's work."

Subsequently, I was troubled by a sense that the smoothness of my tale was partly false. More true was the fact that *The Mirror Dance* was one thing as a book and another as a personal experience. I had wanted to speak of my personal experience, and especially of my difficulties, but in the end I felt I had to justify my experience in order to present it. Thus, I converted my troubles in writing up my study to sources of social scientific insight in the class and published article.

What if, instead, I had produced a less justified personal account that was full of more uncomfortable and unresolved feelings? I had written this type of account in *The Mirror Dance,* using the statements of community members to express ambivalence. But when it came to my own background story, experimentation did not seem in order. My personal feelings seemed to need to be structured in a form that was traditional and clear cut (a success story), or else they would be viewed as self-indulgent, irrelevant, unpublishable. Had I said simply, "Here is some of what I felt in doing this study," I think I would have been more honest. However, I might not have been acceptable.

I think that descriptions of lack of success, lost identity, and unsettled feelings can be valuable in social science, especially when their subject is the social scientist's experience. However, more diffuse descriptions go against the grain because we are used to viewing only certain styles of interpretation as appropriate: those that emphasize the social scientist triumphing over a difficult research experience or a hard problem in writing, for instance, and those that emphasize abstract thinking. In the class on feminist ethnography in which I began to speak about social science writing, the question of how to structure my presentation arose for me immediately. Clearly, the temptation was to play it safe and to organize my tale around success. However, that may not be the best form for a personal account. Similarly, it may not be best to organize an account around an intellectual idea when the subject is one's own experience. For me, it is desirable to structure a description in terms of the emotional content of an experience. The result is a relatively open form that looks more like a mosaic than a linear progression and that communicates on an emotional level rather than as a neat intellectual package.

In the second class to which I was invited to speak about writing, the students had just finished conducting individual research projects. I was supposed to talk about strategies that would help them write up their studies. The subject of the course was organizational communication. In preparation, I made up a list of formats that could be used. When I presented the formats to the class, the students seemed interested primarily in one: a quest form, using a term from Carolyn Heilbrun's *Writing a Woman's Life*.[3] The quest form would allow the students to chronicle a problem of inquiry using their personal dilemmas and research experiences as the central interest of their account. The resulting tale would not be about abstract concepts and their relation to one another, but about each student's search for relevant material, even if this search was frustrated. The students interested in the quest form asked where they could read more about it, both in order to see a good example and so that, if they chose this form for a thesis, they would have a document to wave in some committee member's face which would justify the approach.

At the time, I could give no totally suitable references for what I had in mind, beyond saying there were accounts of the type in anthropology and sociology. A problem, however, was that these accounts tended to be success stories. The story would be one of intellectual triumph rather than of a more unresolved personal quest. Heilbrun's book, discussing quest plots, was not yet published at the time,[4] and, even if it had been, it would not have carried much weight in the eyes of a suspicious department committee composed of social scientists. Heilbrun would be seen as writing about women, biography, autobiography, and fiction, and the assumption was that social science ought to be different. It ought to be more theoretical and abstract.

Later in the class, in response to an item on another list I had drawn up, the students seemed especially interested in discussing their emotions in relation to their studies, particularly some of the emotional difficulties they had experienced doing their research:

disliking certain people they had studied, for instance; feeling awful about continuing; feeling totally frustrated because of not getting useful answers; feeling inseparably part of what they were studying. The students were also concerned about looking toward the future in a positive manner. These were night students studying for their master's degrees. They knew about traditional social science methods, but they had not yet been stifled by those methods or taken them to be unchangeable tenets. I felt that, in their eyes, if an older social science viewed the self as a contaminant, and viewed difficulties in the research process as events to be justified in terms of a pragmatic, positivist rhetoric, a new social science could be different. The books to explain, justify, and legitimate all this had not yet been written, but this class of students was prepared to go ahead as if it had. Their interest in discussing emotions surprised me, as did their relatively open attitude toward what might be included within social science. I was reminded of a comment made by my sister's oldest daughter, who, at age eight, asked her father, a lawyer, why a particular thing could not be done. "Because it's the law," her father said, as one who should know. "Then change the law," the child replied. Clearly, I felt, she had the future on her side.

In the third class, a course on feminist methodology, I was asked to speak about my voice in *The Mirror Dance.* "Where was it?" the students wanted to know, and, "Who was I speaking for?" Again I drew up lists. One list identified six different voices I felt I had in *The Mirror Dance.* I wanted to present the issue of voice as more complex and less controllable than I thought the question suggested. My second list indicated that I felt I was speaking only for myself. In drawing up my second list, I anticipated that the stance of speaking for myself would be unpopular in a feminist class. Feminists were supposed to speak for other people: other women, children, other cultures, the silenced, people who are usually left out. A good feminist, like a good social scientist, was supposed to overcome limits of the self. She should try to give others space to speak in their own words and style.

Some people, I knew, felt I had done that type of work in *The Mirror Dance.* I had written an account that took pains to reflect other people's points of view, using their style of speech. Nonetheless, I now wanted my study to be seen as reflecting my view fundamentally. It was clear to me that my issues organized what I presented as other women's experiences in *The Mirror Dance,* and this seemed the most important feature of the study to point out, especially when others were claiming the opposite. At the time I wrote *The Mirror Dance* and soon after it was published, however, I stressed that my account presented the women of the community speaking. I needed to feel that the issues I described could be found in their lives and that I was being primarily faithful to their reality.

Why the change over time in what I chose to emphasize about my work? To begin with, in looking at my studies and my autobiographical writing, I saw repeatedly that my work looked like itself more than like anything else. It had themes whose source was not in the external world so much as in my consciousness of it. In addition, I found that the people described in my studies usually preferred to have the studies seen as mine and not theirs (see chapter 12). This was also true for people who appeared in stories and novels I wrote. There was more to their stance, I thought, than a desire to deny unwanted truths. There was a basic recognition that a study, or story, was the work of its author: it might include aspects of the lives of other people, but the person most responsible for putting those aspects together would be held accountable for the work in the end. I do not mean here to overlook the possibility of collective, or joint, authorship, but I wish to emphasize that acknowledging one's formative role in constructing a study is important.[5]

Over time, I have come to feel more responsible for my work, or responsible in a different way, than the rhetoric of social science, or of feminism, might suggest.[6] I am responsible for articulating a sense of my own vision and for recognizing that this sense should not be dictated by external prescriptions about what that vision ought to be: that I ought to be representing others, for instance, or that I should view my work as primarily faithful to other realities when it does not seem that way to me. In addition, over time, I have come to feel more caught within myself. I am struck increasingly with the impossibility of getting outside my own skin. The more I try to grasp someone else's experience, the more I am impressed with how hard it is, how much beyond me that other experience really is. This makes me reluctant to present my views as someone else's. I think it is important to try to grasp experiences that are not one's own. However, such attempts ought not to be masqueraded as other than what they are: they are attempts, they grasp only small pieces of experience, and they are always impositions of an authorial perspective.

Long ago, as an undergraduate doing summer field research, I thought, romantically, that when I used other people's words, I was letting them speak through me. Now I am uncomfortable with all denials of the self. Feminism, like social science, and like other ideologies, suggests that the highest good is to be achieved through transcending the self, subordinating it to a more worthy cause or to a common interest.[7] Although I was taught that such a transcendence of the self was desirable, in my experience it has not been possible. I think concepts that help us acknowledge the self can be important in social science. One of these concepts is the idea of individual creative vision. That is what the debates about whose reality is represented in *The Mirror Dance* suggest to me.

The idea that we know the world because we have vision, and each a different vision, is different from the idea that the world is known through selected abstract concepts. Concepts can be detached from the individuals who use them, and passed around, adopted, or discarded. Vision is more complex. It works as a whole and is developed from within.[8] It is less easy to pass around a vision, or to discard it, than to pass around or to discard a concept, or an abstract theory. Instead, one displays, revises, adapts, and changes it. The idea of vision is traditionally more familiar to artists than to social scientists and more associated with a sense of individuality. Erving Goffman, a sociologist, is often said to have had vision, since it is difficult to comprehend his work in a way that fails to acknowledge its uniqueness and its attachment to him. We tend to associate vision with genius, but more accurate would be an association with originality. Total fabrication, or making things up without concern for their relation to the world, is not what social scientists do, but putting pieces together is, and vision can affect how that gets done.

The choice, I think, is whether our individual visions will inform our interpretations in social science, or whether we will ignore our differences in perspective and write accounts that reflect primarily the common terms that are used to describe experience at any time. It is my hope that more choices will be made to draw on individual ways of seeing, and to articulate them fully, no matter how vulnerable the expression of an individual view makes its author. If the social scientific task is to model the world faithfully, there is a need for strategies of interpretation that challenge blinders of conventional thinking, whether these blinders stem from old orthodoxies (e.g., positivism) or new ones (e.g., postmodernist relativism). The self can be used as a source of more truthful expression in the social sciences. However, for this to happen, new forms of socialization may be required. We must teach ourselves that the individual view need not be apologized for and that we have a right to be part of what we know.

6

An Anthropologist
and a Mystery Writer

IN A COLLECTION OF ESSAYS, *Works and Lives,* published in 1988, the anthropologist Clifford Geertz focuses on the "I" of the author in anthropological writing.[1] According to Geertz, up until recently, anthropologists have been seen primarily as fieldworkers rather than as writers. Yet contemporary critiques in social science have called attention to the literary mechanisms of anthropological work and to the way anthropological authority is textually constructed. Geertz's account emphasizes that "all ethnographical descriptions are homemade": they are "the describer's descriptions, not those of the described" (pp. 144–145). In Geertz's view, describers bear a "burden of authorship" (p. 140), which means that those of us who write social science must pay attention to "how words attach to the world, texts to experience, works to lives." We must ask how we know in "other than practical, empiricist terms" (p. 135). The goal is to become increasingly aware of how we create the realities we describe.

A similar subject is dealt with under a different title in Carolyn Heilbrun's 1988 volume, *Writing a Woman's Life.*[2] Where Geertz subheads his book *The Anthropologist as Author,* Heilbrun uses the more pedestrian term "writing," and the particular phrase "a woman's life." Heilbrun is a professor of English, a writer of detective novels, and a feminist. In *Writing a Woman's Life,* she is concerned with limitations on our awareness of female possibilities that result from the nature of the plots we use for thinking about women's lives. The contrast between Heilbrun's views and those of Geertz concerning the relationship between works and lives interests me, not because one of these scholars is right and the other wrong, or because I prefer one to the other, but because of how easily one view (that of Geertz) seems to make the second (that of Heilbrun) unimaginable, or at least not particularly significant. It is as if Geertz—who speaks in a traditional academic voice that comes as if from above, surveying a mottled landscape—is in the foreground. Heilbrun is a shadow figure busy behind him, doing what is traditionally known as woman's work. Her voice comes from the midst of the melee. It is personal, specific, self-effacing. Geertz's voice is authoritative, generalizing, and commanding.

In *Works and Lives,* Geertz makes it a point to distinguish a quality he calls "person-specific" from something else that is called "personal" (p. 6). He emphasizes that anthropological accounts are the former (person-specific), not the latter. The personal, for Geertz, is associated with awkward confessionalism, disease (e.g., the "Diary Disease"), and getting lost in vague mists.[3] The personal is something to be avoided or kept at a distance. Heilbrun, on the other hand, grounds her discussion in an acceptance of the personal: "[Adrienne] Rich asserted here, as she had previously, her belief that it is only the willingness of women to share their 'private and often painful experience' that will enable them to achieve a true description of the world" (p. 68).

Notably, Geertz, in *Works and Lives,* speaks only of biography. For him, the ethnographic task is one of "constructing texts ostensibly scientific out of experiences broadly biographical" (p. 10). Heilbrun speaks of biography and autobiography together, and often alternates between the two as if they were one:

> Catherine Drinker Bowen, the famous biographer of six men, explains how, when asked why she had never written about a woman, she did not dare respond honestly, "I have, six times." She feared, rightly, that she would not be understood. (p. 22)

Heilbrun, in her account, focuses on lives over texts:

> I have chosen to write of women's lives rather than of
> the texts I have been trained to analyze and enjoy.
> (p. 20)

Geertz, however, emphasizes that he deals only with texts:

> If Rabinow in his pages (I am, of course, speaking of
> him and his colleagues only as they function *inside* their
> pages, not as "real persons") . . . (p. 93)

In *Works and Lives,* Geertz wants us to see the artifice at
work in anthropological accounts and to understand how the
first-person singular, implied or directly used, is a way of organizing the consciousness of what appears on the ethnographic
page (pp. 48, 93). To this end, he analyzes the works of four
anthropologists with different approaches to text building. The
first is Lévi-Strauss, for whom, "To reach reality, we must first
repudiate experience" (quoted by Geertz, p. 46). The second
author considered is Evans-Pritchard, in whose work "the over-
riding point of every image, every elegance, every nod, is to
demonstrate that nothing, no matter how singular, resists rea-
soned description" (p. 61). Third is Malinowski, who posed for
ethnographers "a distinctive sort of text-building problem: rendering your account credible through rendering your person so"
(p. 79); finally, there is Benedict, in whose works "the culturally at hand is made odd and arbitrary, the culturally distant,
logical and straightforward" (p. 106).

Geertz draws a distinction between anthropological experience in the field (Being There) and experience on the page (Being
Here). He wants us to look "*at* as well as through" anthropological accounts and to see how they are "made to persuade" within
academic institutions (p. 138). I think his focus on institutions
and on authorial styles is valuable, as are his analyses of specific
anthropological accounts. However, the lives and works he discusses are overwhelmingly male. When Geertz finally takes up
Ruth Benedict, the fourth central figure in his text, the gender
skew in his interpretive stance becomes clear. Benedict stands
alone. She is presented as a token woman who emerged as an

anthropologist at the age of forty and who is sometimes confused with her larger-than-life friend Margaret Mead (pp. 106, 109). Benedict is viewed by Geertz as most interesting when appreciated for her manly qualities or when seen as an atypical woman: "[The] vein of iron in Benedict's work, the determined candor of her style, has not, I think, always been sufficiently appreciated. In part, this is perhaps because she was a woman, and women, even professional women, have not been thought inclined to the mordant" (p. 105). Benedict's "soft-focus lyric poetry" (p. 106), Geertz tells us, and the "onward and upward sermons" with which she began and ended her books (p. 106) are beside the point. Her serious work is neither uplifting nor soft. It is the iron satire, the "incised lines, bitten with finality" (p. 109), the accounts that get a single message across and that are best understood when compared with the works of men: "She did not have Swift's wit, nor the furor of his hatred, and, her cases before her, she did not need his inventiveness. But she had his fixity of purpose and its severity as well" (p. 105).

In reading Geertz on Benedict, I wondered how Geertz, who otherwise sought to be so understanding, could play with Benedict in this way. Why did he use the generic "he" throughout his volume, written, as it was, in the 1980s? Why did he interpret his one central woman as a man? Geertz's chapter on Benedict bears the subtitle "Us/Not-Us," suggesting there may be something significant in the fact that Benedict was a woman, a member of a group socialized to show interest in self-other relationships. Yet anthropologists in general do this, and beyond the title there is little in Geertz's analysis of Benedict that might add to an understanding of how her being a woman affected her work. Geertz's account, in fact, suggests that her gender did not matter, and because of that, I think, Geertz fails to grasp Benedict.

To know Benedict better, one reads Heilbrun, who does not refer to the anthropologist at all in her book. Heilbrun speaks of other women—mostly literary types—and of her own experiences in reading and writing. Where Geertz mentions, in passing, Benedict's quick ascendancy as a mature anthropologist beginning in midlife: her "extraordinarily rapid entry to the discipline—and to the institutional center of it, Columbia's commanding heights" (p. 109), Heilbrun dwells at length on the phenomenon of "women who have awakened to new possibilities in middle age" (p. 124).[4] The difference between the two descriptions is more than one of words. Heilbrun, in discussing what happens specifically to women, is talking in code about the inner life: the nature of personal experience not robbed of its complexity.

Heilbrun uses herself as an example in *Writing a Woman's Life,* offering thoughts that are intended to throw light on other experiences as well. At one point, she discusses her reasons for starting to write novels about the female detective Kate Fansler, under the pen name Amanda Cross:

> When safely hidden behind anonymity, I invented Kate Fansler, I gave her parents, already dead, whom she could freely dislike, and create herself against. . . .
>
> How much of [my] wish to transform female destiny was conscious? None of it. I suspect that if I had been told then that my depriving Kate Fansler of parents indicated any ambivalence on my part toward my parents, I would have disputed the conclusion with vigor. All the conscious reasons for writing were good ones; they operated, they were sufficient to explain my actions. Yet the real reasons permitted me, as other women have found ways to permit themselves, to experience what I would not have had the courage to undertake in full awareness. (pp. 119–120)

Heilbrun then turns, in the same paragraph, to the experience of Virginia Woolf:

> I think Virginia Woolf, for example, early realized, deeply if unconsciously, that the narratives provided for women were insufficient for her needs. . . . All her novels struggle against narrative and the old perceptions of the world. She felt in herself a powerful need for a love we have come to call maternal, a love that few men are able to offer (outside of romance) and that women have been carefully trained not to seek in other women. (p. 120)

Thus autobiography (Heilbrun's Amanda Cross story) merges with biography (Heilbrun's description of Woolf); the work on the page gets entangled in a real life. The boundaries between things are permeable in Heilbrun. They are crossed by means of Heilbrun's emotions, and crossing them is the point.

For Geertz, the drawing of distinctions is what matters. Geertz finds it important to contrast the function of an author with the activities of a writer, for instance. Using Barthes, he suggests that anthropologists are "author-writers" or literary "mules," poised between creating and communicating (p. 20). Heilbrun labels herself a writer only, and one senses she would rather avoid the fuss. If Geertz seems poised between, and above, who he is and what he sees, Heilbrun is repeatedly embedded in it:

> The generation of white women poets I refer to lived through World War II (as I did). Jane Cooper, born in 1924, sets it for us: "World War II was the war I grew up into." (p. 62)

Or,

> Meanwhile, creating Kate Fansler and her quests, I was re-creating myself. Women come to writing, I believe, simultaneous with self-creation. Sand went to Paris and dressed as a boy. Colette was locked in a room by her husband. (p. 117)

Heilbrun appears in her text in small ways: as an assertion within a parenthesis, for instance, or as a brief reference to what she feels she was doing in creating her detective heroine, Kate Fansler. In each case, soon after speaking of herself, she turns her account back outward and speaks of someone else: Jane Cooper, white women poets, women in general, Sand, Colette. Heilbrun's style throughout shows her embeddedness, her wish to be part of her subject.

Geertz, by contrast, speaks from a more distant stance and is less specifically identifiable, either as an individual or in relation to other anthropologists. Traditionally authoritative in his style, he takes authority as his subject in *Works and Lives.* " 'The discourse problem' in anthropology" is, for him, that of "how to author an authoritative presentation"—how to make an account convincing (p. 83). Heilbrun, on the other hand, speaks from an authority that seems often not to be there at all. She is essentially not concerned with authority in *Writing a Woman's Life,* but, instead, with the dynamics of insight. Specifically, she is concerned with how her own insights about plots in the lives of women can be made, at once, to stand out from, and then to merge back with, the insights and experiences of others. Although she makes generalizations in her account, and refers often to the works and words of others, her strongest form of expression is the self-reflective statement:

> Let me return . . . to my two names adopted so cavalierly twenty-five years ago. Something happened both to Carolyn and Amanda: the women's movement. Carolyn began writing in a more personal, for her more courageous, manner, recognizing that women could not speak to other women as men had always spoken, as though from on high. She wrote of herself, she told her own story, she risked exposure. . . . And Amanda became more courageous too: she wrote of feminist matters, and let her heroine continue to smoke and drink, despite frequent protests from readers. . . . For some reason, I was reluctant to reform her. (pp. 121–122)

It is important to Heilbrun's approach that her readers feel the carefulness of her consideration of her own experience. It is on the basis of the quality of her assessment of her experience that her claim to be able to offer something useful to others lies. Her habit of self-scrutiny is very different from the traditional authoritative style, which is what we find in Geertz.

Speaking as an authority "as though from on high," Geertz creates a thick text. It feels deep, weighty, erudite. Geertz catches nuances. He is the master of his subject, distant tutor of all those who would do ethnography. Heilbrun, closer in, has produced a text with a thinner feel. Her goal is clarity more than mastery. Like Geertz, she speaks through an authorial "I" who makes deliberate choices. Most of these choices are not spoken of directly, but they nonetheless pervade her account. Her style suggests a female version of what Becker, in *Writing for Social*

Scientists, describes in one of its male forms: "Some writers—I favor this persona myself—take a Will Rogers line. We are just plain folks who emphasize our similarity to ordinary people, rather than the differences. We may know a few things others don't, but it's nothing special. 'Shucks, you'd of thought the same thing as me if you'd just been there to see what I seen.' "[5]

Heilbrun does not say "shucks." Her presentation of herself in *Writing a Woman's Life* is different, in critical ways, from that of a male common man. She is at once like many others and unlike them, but the differences between herself and others do not threaten her authority so much as they threaten her desire for oneness: her wish to feel together with, and part of, those she writes about. The sense one gets from reading Heilbrun is that for a woman, the presentation of self in a text is less of a game than it is for a man because the woman feels more inextricably bound up with her subject. The task of being a common person differs for a woman because, among women, it is not a hierarchical ordering that counts most, but, rather, mutuality: seeing and being seen, holding and being held.[6]

I am suggesting that female sensibilities affect what women do with themselves in social science writing, and in academic writing generally. Often, women fail to write or speak in traditional authoritative ways, and when this happens, it is important to see what is going on. Perhaps it is not a demonstration of expertise, or a competition. In academic life, the cloak that was supposed to fit never fits the woman as well, and it is necessary to look beneath it.

Works and Lives: The Anthropologist as Author and *Writing a Woman's Life* are thus two different comments on a similar topic: the relationship between a work and its author. These books are affected by gender, whether their authors discuss this topic or not. Both Heilbrun and Geertz are partially hidden in their books. These authors speak about themselves as powerfully in what they do not say as in what they make explicit. In *Works and Lives,* Geertz overlooks gender. Heilbrun's *Writing a Woman's Life* suggests that this omission is significant.

To make their points, both Heilbrun and Geertz speak of others in their books (Heilbrun speaks of literary figures, Geertz of anthropologists), yet in the end, each of these authors is primary. Geertz, at one point, states that individuality is difficult to convey: " 'I' is harder to write than to read," he says, drawing on Barthes (p. 90). However, individuality is central in both these accounts. We wish to tell Geertz apart from other anthropologists; we wish to hear more of Heilbrun's separate voice. *Works and Lives* and *Writing a Woman's Life* are each mysteries. In each, the authorial "I" is sophisticated, well read, and carefully presented. Yet the real "I," more rough, is hinted at. One grasps for a feel of the person behind each study—whether in male tweeds or female folds—and although Heilbrun speaks more directly about herself than Geertz does, in neither case are there enough clues. This is academic prose, as yet uncertain of how much of the self to let in.

III

Individuality

7

Self, Truth, and Form

Lessons from Georgia O'Keeffe

Personal Boundaries

THE PAINTER GEORGIA O'KEEFFE had much to say about the subjective and interpretive role of the self in relation to problems of knowledge. Because, for me, paying close attention to O'Keeffe's words is the best way to grasp her meaning, in this chapter I quote extensively from O'Keeffe's letters, exhibition catalogues, and her one autobiographical volume. I draw on her published writings to further discuss issues that arise in considering the use of the self in social science.

O'Keeffe was extremely straightforward verbally. She seemed to mean exactly what she said, and she defied anyone else to interpret her in a way that was not her own. A good place to begin, I think, is with O'Keeffe's attitude toward words and toward understanding. Generally, O'Keeffe did not like words spoken about her or her art. She especially did not like it when those words were spoken about her by others. She felt that others misinterpreted her. They did not see what she saw or intended them to see:

> I made you take time to look at what I saw and when you took time to really notice my flower you hung all your associations with flowers on my flower and you write about my flower as if I think and see what you think and see of the flower—and I don't.[1]

67

When others attached their meanings to O'Keeffe's work, she felt invaded:

> The little blue book looks so clean and fine—I'll paste things in it but I'm sure—every time I do it—it will shiver and have a queer feeling of being invaded the way I do when I read things about myself.[2]

O'Keeffe's response to invasion was to distance herself. To read any account of her life, no matter how fragmented, is to hear a description of distance, of setting up boundaries, of the artist defining herself as different from others. As Joan Didion writes:

> *The city men. The men. They.* The words crop up again and again as this astonishingly aggressive woman tells us what was on her mind when she was making her astonishingly aggressive paintings. It was those city men who stood accused of sentimentalizing her flowers: "I made you take time to look at what I saw and when you took time . . . you hung all your associations. . . ." *And I don't.* Imagine those words spoken and the sound you hear is *don't tread on me.* . . . The men yearned toward Europe so she went to Texas, and then New Mexico.[3]

A sense of setting up boundaries appears in the account of O'Keeffe biographer Laurie Lisle:

> The world out of which O'Keeffe's art emerged contained no imperfect, boring, bothersome human beings. Henry McBride jokingly suggested in a review he wrote in 1941 that when she painted her southwestern "heaven," Georgia should paint some steel bars across the blue sky to keep any intruders out.[4]

The name Georgia O'Keeffe has come to mean going off, getting far away from, living in isolation in order to live. The fact that O'Keeffe painted in a way that others did not paint—bigger, brighter, differently, of clam shells and colors—has become less important than the fact of where she painted: the remove she insisted upon as necessary for her work. When she spoke of settling in New Mexico, O'Keeffe was also speaking of an internal state:

> I must be someplace where the people do not run me crazy. . . . I have heard from various sources that I was going west. (*A&L*, p. 221)

> I am about 100 miles from the railroad—68 from Santa Fe—95 from Taos—40 miles from town—18 miles from a post office. (*A&L*, p. 230)

> The things I like here seem so far away from what I vaguely remember of New York. (*A&L*, p. 230)

> I have bought a house out here. (*A&L*, p. 230)

O'Keeffe painted in a place where the ideas of other people could be kept off and her own ways nourished. Yet this was an accomplishment, and it was the result of many internal recognitions. For example, it was important to O'Keeffe to acknowledge the critical role of having a standard of judgment of her own, rather than working to please others:

> It is curious—how one works for flattery—Rather it is curious how hard it seems to be for me right now not to cater to someone when I work—rather than just to express myself. (*A&L*, p. 145)

> I decided I was a very stupid fool not to at least paint as I wanted to. (*GOK*, opposite plate 12)

> One thing that gets me about . . . the Taos country— it is so beautiful—and so poisonous—the only way to live in it is to strictly mind your own business—your own . . . your own pleasures—and use your ears as little as possible. (*A&L*, p. 201)

> My center does not come from my mind—it feels in me like a plot of warm moist well tilled earth with the sun shining hot on it. . . . It seems I would rather feel it

starkly empty than let anything be planted that can not be tended to the fullest possibility of its growth. . . . If the past year or two or three has taught me anything it is that my plot of earth must be tended with absurd care— By myself first. (*A&L*, p. 217)

O'Keeffe was often alone:

The days here are good. They are mostly days alone. (*A&L*, p. 203)

I get out my work and have a show for myself before I have it publicly. I make up my own mind about it— how good or bad or indifferent it is. After that the critics can write what they please. I have already settled it for myself so flattery and criticism go down the same drain and I am quite free. (*GOK*, opposite plate 31)

I suppose the moments one most enjoys are moments alone—when one unexpectedly stretches something inside you that needs stretching. (*A&L*, p. 199)

In part because it is singular and direct, O'Keeffe's voice is different from many we are used to. Didion says:

On the evidence of her work and what she has said about it, Georgia O'Keeffe is neither "crusty" nor eccentric. She is simply hard, a straight shooter, a woman clean of received wisdom and open to what she sees. This is a woman who could early on dismiss most of her contemporaries as "dreamy," and would later single out one she liked as "a very poor painter." (And then add, apparently by way of softening the judgment: "I guess he wasn't a painter at all. He had no courage and I believe that to create one's own world in any of the arts takes courage.") This is a woman who in 1939 could advise her admirers that they were missing her point.[5]

I like O'Keeffe's resoluteness, and her paintings, and I am interested in her words, although I have difficulty understanding how she could live as apart from other people as she did. Georgia O'Keeffe has long been part of my inner landscape, which means that when I see her paintings or read her words, I listen as if her voice were some version of my own. I also listen without being entirely clear that her voice is not my own.[6]

Excerpts from O'Keeffe's Letters

In November 1987, eight months after Georgia O'Keeffe's death at the age of ninety-eight, a retrospective exhibit of 120 of her works opened in Washington, D.C., the first stop in a tour of five U.S. cities. The catalogue accompanying this retrospective exhibit was a thick art book, *Georgia O'Keeffe: Art and Letters,* containing prints of the pictures shown and copies of 125 letters written by O'Keeffe, most of them previously unpublished. Initially, I did not expect to see the O'Keeffe retrospective. I read the letters instead, thinking that reading the letters and looking at the prints in the catalogue was a preferable experience to attending the show. At the exhibit, I thought, I would only get upset. I would feel that the exhibit was a memorial service for O'Keeffe, and I would remember too acutely that she was dead. As I read the O'Keeffe letters in the privacy of my home, in the context not only of O'Keeffe's work but of my own life, I thought about how O'Keeffe's statements might be relevant to social science.

In her letters, O'Keeffe's comments about self, truth, and form stood out for me. Her remarks on this topic were few, but they seemed to say a great deal. The following fourteen excerpts, arranged chronologically, suggest the nature of O'Keeffe's views. These statements suggest a different kind of orientation than we are used to in social science, one that emphasizes the particularity of individual vision and that views self and work as one. It is a view that takes as central the task of "making the unknown known" through the creation of form. It is also an orientation that focuses on the effort of an artist to speak from within the self.[7]

To Anita Pollitzer, 4 January 1916 (A&L, p. 147)

Dear Anita:

I wonder what I said—I wonder if any of you got what I tried to say—Isn't it damnable that I can't talk to you If Stieglitz says any more about them [O'Keeffe's charcoals]—ask him why he liked them—

To Alfred Stieglitz, 1 February 1916 (A&L, p. 150)

Mr. Stieglitz:

—Words and I are not good friends at all except with some people—when I'm close to them and can feel as well as hear their response—I have to say it someway—Last year I went color mad—but Ive almost hated to think of color since the fall went—Ive been slaving on the violin—trying to make that talk—I wish I could tell you some of the things Ive wanted to say as I felt them. . . . I am glad they [the drawings] said something to you.—I think so much alone—work alone—am so much alone—but for letters—that I am not always sure that Im thinking straight—

To Paul Strand, 25 June 1917 (A&L, p. 163)

—I want—just to reach out my hand to you and let you hold it—Can you understand that—its different from telling you in words what they [Strand's photographs] say to me—in a way it is much more real—maybe thats why I want to touch people so often—its only another way of talking

From the same letter to Paul Strand (A&L, p. 163)

You see—when I saw the large prints—they made me see you better—what was in the prints must be in you—then it is a marvelous discovery—that that could be in anyone now—feeling more definitely what is in you—what you are—your songs—that I see here—are sad—very wonderful music—quieting music of the man I found a real living human being—and the man I found in the big prints—You would not be what you are to me at all if I had not seen the part of you that is in the big prints—These little ones are merely songs that the man I have come to know is singing

To Paul Strand, 24 October 1917 (A&L, p. 165)

Paul—

Love you both armsful—Or is it what you say that I love—

To Sherwood Anderson, 1 August 1923? (A&L, p. 172)

Dear Sherwood Anderson:

. . . Ive wanted so often to write to you—two things in particular to tell you—but I do not write—I do not write to anyone—maybe I do not like telling myself to people—and writing means that.

To Sherwood Anderson, September 1923? (A&L, p. 174)

Ive been thinking of what you say about form—. And either I dont agree with you—or I use a different way of thinking about it to myself—Maybe we mean somewhat the same thing and have different ways of saying it—

I feel that a real living form is the natural result of the individuals effort to create the living thing out of the adventure of his spirit into the unknown—where it has experienced something—felt something—it has not understood—and from that experience comes the desire to make the unknown—known—By unknown—I mean the thing that means so much to the person that he wants to put it down—clarify something he feels but does not clearly understand—sometimes he partially knows why—sometimes he doesn't—sometimes it is all working in the dark—but a working that must be done—Making the unknown—known—in terms of one's medium is all absorbing—if you stop to think of form—as form you are lost—The artists form must be inevitable—You mustn't even think you wont succeed—

From the same letter to Sherwood Anderson (A&L, pp. 174–175)

—Making your unknown known is the important thing—and keeping the unknown always beyond you—catching crystalizing your simpler clearer vision of life—only to see it turn stale compared to what you vaguely feel ahead—that you must always keep working to grasp—the form *must* take care of its self if you can keep your vision clear—

You and I dont know whether our vision is clear in relation to our time or not—No matter what failure or success we may have—we will not know—But we can keep our integrity—according to our own sense of balance with the world and that creates our form—

What others have called form has nothing to do with our form—I want to create my own I cant do anything else—if I stop to think of what others—authorities or the public—or anyone—would say of my form I'd not be able to do anything

I can never show what I am working on without being stopped—whether it is liked or disliked I am affected in the same way—sort of paralyzed—.

To Mabel Dodge Luhan, 1925? (A&L, p. 180)

Mabel Luhan:

Last summer when I read what you wrote about Katherine Cornell [the actress] I told Stieglitz I wished you had seen my work—that I thought you could write something about me that the men cant—

What I want written—I do not know—I have no definite idea of what it should be—but a woman who has lived many things and who sees lines and colors as an expression of living—might say something that a man can't—I feel there is something unexplored about woman that only a woman can explore—Men have done all they can about it.—Does that mean anything to you—or doesn't it?

Do you think maybe that is a notion I have picked up—or made up—or just like to imagine? Greetings from us both— And kiss the sky for me—

To Waldo Frank, Summer 1926 (A&L, p. 184)

I have just read the article on the Jew in the Menorah that you sent Stieglitz

I was much interested—Maybe not as much in the article as in something I feel you trying to do with yourself—

To William M. Milliken, 1 November 1930 (A&L, p. 202)

Dear Mr. Milliken:

I know I can not paint a flower. I can not paint the sun on the desert on a bright summer morning but maybe in terms of paint color I can convey to you my experience of the flower or the experience that makes the flower of significance to me at that particular time.

Color is one of the great things in the world that makes life worth living to me and as I have come to think of painting it is my effort to create an equivalent with paint color for the world— life as I see it

To Dorothy Brett, ? September 1932 (A&L, p. 210)

Dear Brett:

About your Lawrence book—I feel so surely that if it is to be of any value to anyone—it must be your truth as nearly as you are able to put it down. It seems to me the only truth of him that you can create—You must write it truly to the end for yourself— Then if you do not wish to print it to the end stop where you wish—My personal feeling about it is that there is no point in doing it if it is not a truth—Ones truth must always have a certain amount of fantasy in it—I think the human being func- tions that way—but the fantasy must be one's truth to keep ones whole consistent—It must be your real experience—you cant make up anything as good as what happened if the whole thing was as wonderful [for] you as you have made me feel it was

To Cady Wells, 3? February 1938 (A&L, p. 223)

Dear Cady:

When Stieglitz looked at the water colors [sent by Wells] he remarked almost what I had said—they will be liked—they will sell—they are good—very good—but they would not make a show at the Place.

You see Walker [a New Mexico painter] was not a good painter—he seldom had anything that one felt was complete but he was a searcher. He moved toward something very big and simple. I always felt he would not live long enough to work through to anything of importance—but I liked his direction—his intelligence and a kind of truth that he had—His self was always in his painting—and may I say to you without offending—that I do not find your self in your paintings—there is beautiful taste

With your mind you can organize a well balanced picture—and the color is always good—very good . . . —but some way the thing that makes a painting live and breathe isn't there for me and I don't know why—

To John Baur, 22 April 1957 (A&L, p. 267)

From experiences of one kind or another shapes and colors come to me very clearly—Sometimes I start in a very realistic fashion and as I go on from one painting after another of the same thing it becomes simplified till it can be nothing but abstract—but for me it is my reason for painting it I suppose

At the moment I am very annoyed. —I have the shapes—on yellow scratch paper—in my mind over a year—and I cannot see the color for them—Ive drawn them again—and again—it is something I have heard again and again till I hear it in the wind—but I can not get the color for it—only shapes—None of this makes sense—but no matter

Comments on O'Keeffe's Perspective

In the preceding excerpts from her letters, O'Keeffe looks repeatedly for the artist's self in her work: "May I say to you without offending—that I do not find your self in your paintings . . . the thing that makes a painting live and breathe isn't there for me" (*A&L*, p. 223). She views the self in a work as active: "I was much interested—Maybe not as much in the article as in something I feel you trying to do with yourself" (*A&L*, p. 184).

For O'Keeffe, the creation of a work of art is a process of knowing: "making the unknown known" through the invention of form (*A&L*, p. 174). To be of value, an invented form must be a statement of truth, not abstract or general truth, but particular truth, sourced in the specific experience of a knower: "I feel so surely that if it is to be of any value to anyone—it must be your truth as nearly as you are able to put it down" (*A&L* p. 210).

I think that O'Keeffe's concern for the presence of an author's self in a work, and her sense of viewing a work as part of a self in progress, provides a contrast to contemporary critical views of literature and art. The latter underplay the role of an original author and, instead, call attention to the interpretive activity of readers who assign their own meanings to others' works.[8] According to contemporary (postmodern) theory, anyone can assign a meaning with almost equal validity. The artist's meaning is not privileged. According to O'Keeffe, some meanings belong to others and are not hers, and there is not an equivalence between them. Meanings attached to her work by others often violate her own sense of what her painting is about. When this happens, it is her word we should listen to because her work is her effort to speak and she wants to be understood:

> I wonder if any of you got what I tried to say. (*A&L*, p. 149)

> I have painted portraits that to me are almost photographic. I remember hesitating to show the paintings, they looked so real to me. But they passed into the world as abstractions—no one seeing what they are. (*GOK*, opposite plate 55)

So I said to myself—I'll paint what I see. . . . I will make even busy New Yorkers take time to see what I see of flowers. (*GOK*, opposite plate 23)

I don't mind if you say it is a tiresome painting—that is a matter of personal opinion—but it is not an enlarged flower. (*Portrait*, p. 179)

I am often amazed at the spoken and written word telling me what I have painted. (*GOK*, p. 1)

The things they write sound so strange and far removed from what I feel of myself. (*Portrait*, p. 179)

They were only writing their autobiography . . . it really was not about me at all. (*A&L*, p. 171)

They were talking about themselves, not about me— the people that saw them that way.[9]

One day I saw a man looking around at my Halpert showing. I heard him remark, "They must be rivers seen from the air." I was pleased that someone had seen what I saw. (*GOK*, opposite plate 103)

In addition to claiming a primacy for her own understanding of her work, an understanding often at odds with the views of others, O'Keeffe maintained a clear subjectivist stance:

I know I can not paint a flower. I can not paint the sun on the desert on a bright summer morning but maybe in terms of paint color I can convey to you my experience of the flower or the experience that makes the flower of significance to me at that particular time. (*A&L*, p. 202)

I was on a stretcher in a large room, two nurses hovering over me, a very large bright skylight above me. I had decided to be conscious as long as possible. I heard the doctor washing his hands. The skylight began to whirl and slowly become smaller and smaller in the black space. I lifted my right arm overhead and dropped it. As the skylight became a small white dot in a black room, I lifted my left arm over my head. As it started to drop and the white dot became very small, I was gone. A few weeks later all this became the black abstraction. (*GOK*, opposite plate 54)

I find that I have painted my life—things happening in my life—without knowing. After painting the shell and shingle many times, I did a misty landscape of the mountain across the lake, and the mountain became the shape of the shingle—the mountain I saw out my window, the shingle on the table in my room. I did not notice that they were alike for a long time after they were painted. (*GOK*, opposite plate 52)

A little way out beyond my kitchen window at the Ranch is a V shape in the hills. I passed the V many times—sometimes stopping to look as it spoke to me quietly. I one day carried my canvas out and made a drawing of it. (*GOK*, opposite plate 85)

I long ago came to the conclusion that even if I could put down accurately the thing that I saw or enjoyed, it would not give the observer the kind of feeling it gave me. I had to create an equivalent for what I felt about what I was looking at—not copy it. (*GOK*, opposite plate 63)

This is a drawing of something I never saw except in the drawing. When one begins to wander around in one's own thoughts and half-thoughts what one sees is often surprising.[10]

I want to paint in terms of my own thinking and feeling. (*Portrait*, p. 162)

Thus despite the externally representative quality of her paintings, O'Keeffe was depicting an inner world: how she felt about mountains, the significance of a flower "to me at that particular time." Her forms of expression—paint shapes and colors on canvas—were, for her, forms of knowledge. They were also nonverbal:

> From experiences of one kind or another shapes and colors come to me very clearly. (*A&L*, p. 267)

> Words and I are not good friends at all. (*A&L*, p. 150)

> Such things don't make sense except painting sense. (*Portrait*, p. 179)

> The feeling that a person gives me that I can not say in words comes in colors and shapes. (*A&L*, p. 218)

> A hill or tree cannot make a good painting just because it is a hill or a tree. It is lines and colors put together so that they say something. For me that is often the most definite form for the intangible thing in myself I can only clarify in paint. (*GOK*, opposite plate 88)

> The meaning of a word—to me—is not as exact as the meaning of a color. (*GOK*, p. 1)

Finally, O'Keeffe was concerned with beauty:

> An idea that seemed to me to be of use to everyone— whether you think about it consciously or not—the idea of filling a space in a beautiful way. (*GOK*, opposite plate 11)

> I covered the canvas as best I could and thought the trees particularly beautiful. (*GOK*, p. 13)

> All the earth colors of the painter's palette are out there in the miles of badlands. The light Naples yellow through the ochres—orange and red and purple

earth—even the soft earth greens. You have no associations with those hills—our waste land—I think our most beautiful country. (*GOK,* opposite plate 26)

It is a very beautiful world—I wish you could see it. (*A&L,* p. 233)

In her day, O'Keeffe was seen as expressing the sensibility of "woman." By this, it was meant that she was sensual, intuitive, and erotic. However, she rejected such labeling fiercely. I think she would also reject attempts to categorize her subjective stance as one-half of human experience, the other half being a rational or objective approach. For her, subjectivity was all encompassing, the only thing of real value, the source of creative intelligence. Although familiar in the arts, within social science, subjectivity is acknowledged with difficulty. In social science studies, statements that reflect individual experiences are often accompanied by statements intended to counter the distorting effects of a personal view. O'Keeffe's comments about the process of her painting suggest that we enter a world in which the subjective individual view is not countered, however, but rather accepted as defining a world. In that world, truth is found in the effort to articulate an inner and individual reality. O'Keeffe's stance is a difficult one to accept if one has been taught that subjective individual knowledge is untrustworthy or secondary, that it is the lens through which one views the world, and not the world itself.

Yet O'Keeffe says, look at truth differently. See it as individual and internal. She suggests a different order of ideas about what we might strive for than the set of ideas normally emphasized when we study social science method. The idea of striving to speak from within, to make one's particular unknown known, is different from the idea of striving to speak about something outside the self. O'Keeffe believes that the mountains she paints exist in the desert landscape aside from her, but when she paints, she grasps not for those external mountains, but for her view of them, and the difference between the two kinds of efforts is important. Usually in social science we think of ourselves as describing an outer world. The definition of our effort as that of grasping an external reality affects our vocabulary and probably the nature of the pictures we draw about our experience. O'Keeffe's writings suggest that there may be something useful to be learned from turning the entire perspective around.[11]

Finally, O'Keeffe's stance is notable for its clarity concerning compromise in the individual's depiction of truth. In contrast to contemporary views that stress the importance of a negotiated reality (that a study ought to be a product of negotiations between researcher and researched, for instance), O'Keeffe was not interested in striking bargains with anyone concerning what she might express or know. She was interested in grasping for herself what her truth was to be: "It is something that I have heard again and again till I hear it in the wind—but I cannot get the color for it" (*A&L,* p. 267). For O'Keeffe, in addition, the social scientific self-other dichotomy was not relevant. The issue was how to speak: "I wish I could tell you some of the things Ive wanted to say as I felt them" (*A&L,* p. 150).

In many discussions of method in social science, there is a concern with distinguishing between self and other, observer and observed, the view of the native and of the stranger, and with reconciling two such perspectives. The central task is seen as one of matching a researcher's, or theorist's, concepts to the world, using the self to describe what is not the self. O'Keeffe's perspective, in my view, denies the need for such a self-other dichotomy and proposes a reframing of what is going on. O'Keeffe's discussion of her art is repeatedly a discussion of an attempt to speak[12] in which the speaker does not use herself to describe an outer world, but, rather, to articulate the world of the self. O'Keeffe seeks to discover not the nature of an "other," but to find out what she wants to say, from within, and to determine how to say it: to find the color to express a feeling, for example. In her attempt to speak, O'Keeffe's self and world are not separate. The self is the world, and to deny the self is to deny the world. This orientation, I think, is important. It is usually not prominent, and not viewed positively, in the thinking of social science. A perspective such as O'Keeffe's, which is centrally concerned with individual expression, is assumed to be more appropriate to the world of the arts, or it is seen as too egotistical, or too inward-looking, for social science.

For me, O'Keeffe's perspective is liberating. I identify a

great deal with O'Keeffe's attempt to grasp shapes and meanings in the darkness of her own mind. I identify with her desire to say, "this is what something is to me at this particular time." I identify in these ways even as I find myself offering a description intended also to reflect an outer world (a lesbian community, or a social science ideology). I think O'Keeffe is important to me in part because most of the social scientific community, and most of the larger world that I know, tells me to accommodate, to reflect the other, compromise, forget myself, pretend, agree, use the words that people generally use, seek for objectivity, or for what many might think is true. O'Keeffe is a special figure in my inner landscape because I hear her advising me to pay attention to an inner struggle and to acknowledge that the hardest part of my effort—in social science and in other writing—is something I feel within the confines of myself. The most difficult work is to make my inner experience into material that can be drawn from, that can be used as the basis for an outer picture or story.

Others may have their own O'Keeffes, I imagine: people who help them to clarify the nature of their efforts and who help them to talk about how their individual processes may be different from what is commonly done around them. In the midst of all the contemporary debates about whose approach to truth one should use—the positivist's or the relativist's, the feminist's, the Marxist's—I know I am grateful for O'Keeffe. I am glad for her high-mindedness, her arrogance, her individuality, her clarity. I am glad for her emphasis on articulating her own truth and on conveying the emotional meaning that mountains and flowers have for her. However odd it may sound to assert that social science is about what can be found in the self, that is what I wish to say above all. I wish to do this in part because the nature of my effort feels that way to me: I do not struggle to get the outside world to speak, but to find my own version, my own reality. In part, I emphasize the self because O'Keeffe, or someone like her, encourages me, and, in so doing, relieves me of a burden. It is a great relief for me to imagine O'Keeffe, as she appears in a Stieglitz or Adams photograph: resolute, intolerant of all challenges to her art, disdainful of those who would refute her. She stands as a woman refusing to be a woman in many of the ways expected of her, yet in her defiance, seeking solutions to a very female predicament: she speaks of a need to bound the self against violation, at the same time as articulating a desire for a continuity of self with other. Her continuity is that of a painter with a landscape, of subjectivity with truth, of an artist with her paintings. Perhaps I am not like O'Keeffe at all. I do not paint; I write. My sense of clarity about myself is often missing. My contributions are far smaller than O'Keeffe's. I am not like her, but I certainly wish to be like this woman who refused to let the world define her meaning.

The unity of self, art, and world, the creation of form through processes of individual knowing, and the central nature of a subjectivist stance are some of the themes that emerge from Georgia O'Keeffe's writings.[13] O'Keeffe's verbal statements are remarkably consistent, whether found in letters published after her death, in exhibition catalogues, or in books appearing in her lifetime. To me, O'Keeffe's views are useful in social science not because they are the views of an artist, but because they convey the perceptions of an unusually astute woman, because they deal fundamentally with issues of how we know, and because they articulate an alternative to the depersonalized and consensual positivism that has historically been seen as necessary to the conduct of science. They are also a source of inspiration: "I said to myself, 'I have things in my head that are not like what anyone has taught me—shapes and ideas so near to me—so natural to my way of being and thinking that it hasn't occurred to me to put them down.' I decided to start anew—to strip away what I had been taught—to accept as true my own thinking. This was one of the best times of my life" (*GOK*, opposite plate 1).

8

From O'Keeffe
to Pueblo Potters

I HAVE JUST RETURNED from a trip to New Mexico. I did not go in search of Georgia O'Keeffe, but I did see a small exhibit of O'Keeffe paintings while in Santa Fe. In the Museum of New Mexico, one of the O'Keeffe paintings on display was *Turkey Feathers and Indian Pot,* a portrait of a shining black pot with gray-black feathers standing in it. The canvas showed O'Keeffe's virtuosity: she could paint a picture of a pot that seemed more desirable than the real thing. The *Feathers and Pot* portrait was forty-seven years old. It had been saved by a private collector and loaned to the museum for this small show. The larger O'Keeffe retrospective, on tour at the same time, contained 120 pictures on many more walls. It was now in Dallas and it would soon be in New York.[1]

I stood in the museum in Santa Fe in the room where about twelve O'Keeffes hung, and I could not stay long enough. "Her actual brush strokes," I thought, never having seen an original O'Keeffe before. On some of the paintings, individual brush strokes could not be seen: the paint was laid on in a style that smoothed them away.[2] On other paintings, one or only a few brush strokes were visible, and these appeared deliberately exposed, as if the artist was using them to identify her work as a painting. There was a single brush stroke in the belly of the black pot; a picture of a river backed by mountains showed many

85

strokes. In looking at these brush strokes, I suddenly felt that I had met O'Keeffe. She was standing not far off, viewing a scene, gauging perspective, applying paint.

She was reminding me that art is selective: one tells and does not. O'Keeffe could have painted realistically: she had the talent, just as the novelist Joan Didion could have written about weather when she chose not to ("I happen to like weather, but weather is easy").[3] O'Keeffe painted, instead, what was important to her. Through her paintings, she grasped feeling and made an effort to speak. If O'Keeffe's words are to be believed, what others continue to see in her work is probably not what she saw and not what she intended us to see. That her pictures hang on our walls, or are stored in our closets, indicates that something is valued, but it is not a shared insight or a similar labor. O'Keeffe did not sign her paintings. When asked why not, she is said to have answered, "I might as well sign my face."[4] That similar statements are rare in social science, and in painting generally, says a great deal about the extent to which we reward conformity.

I used to live in New Mexico. At the time, I knew that O'Keeffe lived in the same state an hour and a half away, but I did not think much about it. I had moved to the Southwest to teach at a state university. I stayed for two years and, since then, have gone back three times, but more often in my mind. The scenes I return to are weighted with emotion: full of memories of people, friendships, and romances, and camping trips to odd places, mostly alone. When I imagine these scenes, I see stretches of desert, brown mountains, trees by a river, wildflowers, adobe-style homes. When I lived in New Mexico, I had a few friends who were interested in local crafts and who were glad when I first bought a handwoven wool rug. Although made in Mexico (the country), the rug seemed to suggest that I was no longer treating myself as beyond attachments: here today, gone tomorrow, traveling light. I was willing to carry some extra baggage.

Starting with that rug, I began to develop a small collection of New Mexico crafts: not expensive things and not many of them, but enough to remember with. Among my new possessions were four Pueblo Indian pots. One evening in a rage two years ago, I broke these pots, fighting someone (I am not sure who), seeking to expunge a period of my past, precisely because that was something I could not do. The pots had been sitting on a bookshelf in my study in a house where I now lived in San Francisco. I used to look at them while I worked. After breaking them, I felt a great loss. I then found three pieces of California pottery and put them on my bookshelf in the empty space left by the absence of the Indian pottery. The new pottery seemed less fragile: a white casserole, a pitcher, a bowl. Yet images of the first set of pots persisted. I saw them sitting across from me still, although they were physically gone.

On my recent trip to New Mexico, I looked again at Pueblo pots. I wanted to replace the ones I had lost with others like them. Toward the end of the trip, on a paper placemat in an Albuquerque restaurant, I wrote these words: "emotion, detail-work, balance (although not necessarily symmetry), and uniqueness." I was trying to describe why a pot appealed to me, but describing it in terms of abstract qualities of the pot, rather than in terms of specific experiences of my own that led me to like certain pots and find them meaningful. On my list, first, I noted that a pot had to appeal emotionally. It had to make a statement of feeling—as might a painting, a piece of social science, or a part of a desert landscape framed with the mind's eye. What I did not note was that someone I loved had once given me a pot while I lived in New Mexico; that my friends there had valued Pueblo pottery; that the first woman I ever lived with used to put out two pots made by a friend of hers each time she and I moved to a new house. We moved often, and this ritual helped us to make each new place feel like home.

Second on my list was the statement that a pot drew one close through the nature of its detail, whether this detail was a design painted on the surface of the pot or a carefully polished finish. Balance, the third term, pointed to the relation between the parts: a narrow mouth leading to a wide belly, for instance, or a totally rounded shape. Uniqueness meant one of a kind.

Perhaps predictably, although it puzzled me afterward, on rereading my list of pot qualities, I felt it was not interesting. I lengthened the list and then shortened it to improve it, but it still was not compelling. Why did I make up this list to begin with? I think I wanted to convert my personal experience into a more important statement. Thus I took my immediate and somewhat difficult experience of looking for things to capture my life with, and I hid it in a list of general attributes of pottery. I thought that doing this would help me make a statement that would be viewed as significant in social science, that would address questions of how a piece of art works: what makes it meaningful, powerful, and valuable. If I could answer the art question, I might also be able to say something about how a social science study works.

However, in the process of making up my list of abstract qualities of pottery, my specific personal experiences got lost, or simply disappeared. One of these experiences was that as I looked at the Pueblo pots in New Mexico, I kept looking for pots that were different. I wished to imagine a few Indian potters sitting at home, or where they worked, with their sights set on making clay vessels that were unlike others—that were too different to sell in the market or to be accepted as representative of their kind. This wish probably reflected not a vision of some other woman sitting faced with her day's clay, but my thinking about my own work. It spoke of a desire not as grand as the ambition to do what has not been done before, but no less significant: the desire to articulate a sense of oneself. Ursula LeGuin, the science fiction writer, has written of this desire in an essay on narrative:

> Why do we tell tales, or tales about tales—why do we bear witness, true or false? . . . Is it because we are so organized as to take actions that prevent our dissolution into the surroundings? I know a very short story which might illustrate this hypothesis. You will find it carved into a stone about three feet up from the floor of the north transept of Carlisle Cathedral in the north of England. . . . Here is the whole story: *Tolfink carved these runes in this stone.*[5]

LeGuin is speaking about a man whose marks in stone were important because they were his. To consider Pueblo pottery is to deal with people in another culture and a different time. Without making any universalistic claims about a human need to leave one's mark, I wish to suggest that a desire to differentiate oneself from one's surroundings, both social and physical, can be found in very different settings, as can a desire to imbue one's work with a sense of one's self.

After my trip to New Mexico, I started reading the words of Pueblo potters:

> No two potters are going to make the same pottery. A part of their personality goes into the pot. (Hopi)[6]

> So much of me goes into the pot. (Laguna)[7]

> Even my thoughts are in the pot. If you're angry and you are making pots with bitter feeling toward others or toward something, your pot will act accordingly. (San Ildefonso)

> When you lose a pot, you lose yourself. (Taos)

> I have a hard time parting with my pieces; it's part of me that's going too. Whatever you have put into that piece has helped make you a better person. When you are done and you hold it in your hands, to me, it comes alive. (Jemez)

> My pots start out somewhere deep inside me. I feel that physically I just make what comes out of me spiritually. (Santa Clara)

In these statements, the line between maker and made is unclear. Rather than an orientation that says an artist can be detached from her work, these potters assert a vision in which what happens to the pot happens to the potter. The pot carries its maker's thoughts, feelings, and spirit. To overlook this fact is to miss a crucial truth, whether in clay, story, or science. I am speaking of a need for connection. To ignore the continuity between maker and made is to describe a world of objects where the individual is not seen, where the presence of an artist is not recognized in her work, the presence of a scientist not acknowledged in a study. The world of creative endeavor thus becomes disjointed, and those who do its labor become alienated.

Georgia O'Keeffe asserted a sense of herself in her painting and nourished her sense of self by living apart from others and by painting in a distinctive style. The Pueblo potters quoted here nourish a sense of themselves in a different way: by emphasizing their ties with others and by producing pots in traditional Pueblo styles. Yet, despite their differences, these potters share with O'Keeffe strong feelings about the importance of an involvement of self in work. That is why their statements interest me. Social science studies, like Pueblo Indian pots and O'Keeffe paintings, have little meaning aside from that given to them by those who grant these works a place in their lives. The artists who make pots and paintings invest themselves in their works, as do those who, in appreciating their art, find a place for its spirit within them:

> When someone else likes the piece, you feel really good. I like to sell to collectors because you know that they are buying it for their own self and they will hoard the pot. It's a different feeling than selling to galleries or wholesalers. (A Taos potter)

> Perhaps more than anything, it comes down to the way the pottery *feels*—a feeling that has to do with the potter's intent. All ethical potters agree that they must be honest, that they must tell the truth about their materials. (A journalist)

> It's a very exciting kind of feeling . . . to be able to pick up a piece and say I love this piece. I'm going to take it home. (A Santa Clara potter)[8]

In social science, we emphasize thinking: the self is largely seen as an intellectual construct, the seat of cognition. Our studies are also viewed as functioning primarily on an intellectual level—in terms of how they structure thought. We tend to overlook emotions. Yet emotions bridge gaps; they enable connections between self and work: "I enjoy the painting because I paint how I feel" (a Hopi-Tewa potter).[9] Why is it so hard to say, similarly, of our studies, I write how I feel?

9

Pueblo Indian Potters

Individual Difference in a Collective Tradition

Reyes had always taught her daughters that it was the woman's part of living to hold things together. Men could build up or tear down houses and ditch banks; but women put clay and sand together to make pottery, or cooked several foods at one time to make one dish. That was part of a woman's life, to make things whole.—Alice Marriott, María: The Potter of San Ildefonso

From the foot of Enchanted Mesa one can see the Acoma pueblo mesa beyond. . . . The top of the rock, seen from this distance, undulates with the rhythm of the different rooftops. The adobe covering the walls of the houses comes from the surrounding land. . . . If one did not know there was a village here, it would be easy to miss.—Susan Peterson, Lucy M. Lewis: American Indian Potter

Part I: Bunzel's *Pueblo Potter*

IN 1929, the anthropologist Ruth Bunzel published a study of New Mexico Pueblo Indian potters based on fieldwork conducted in 1924–25. *The Pueblo Potter: A Study of Contemporary Imagination in Primitive Art* is, according to Bunzel, a study especially of the manner in which an individual operates within the limits of an established style, or finding that impossible, creates new values and wins for them social recognition. It is an attempt to enter fully into the mind of primitive artists; to see their technique and style, not as they appear objectively to students of museum collections, but as they appear to the artists themselves, who are seeking in this field of behavior a satisfactory and intelligible technique of individual expression. [1]

Its references to primitivism notwithstanding, Bunzel's study interests me because of its concern with individual expression. The Pueblo cultures in which Bunzel studied pottery making placed a great value on collective life and on the unity of self and environment. Bunzel does not dwell on the way individuality is therefore different in the Pueblo context than it is elsewhere, [2] but her account is instructive in other ways. In focusing on "decorative problems in the manufacture of household articles of clay" (p. 2), Bunzel attends to work done primarily by women. In taking as her topic the phenomenon of individual expression, she deals with the attempts of Pueblo potters to leave their mark in an otherwise highly conventionalized art form.

Pottery Making and the Puzzle of Individuality

Bunzel begins *The Pueblo Potter* by describing the making of Pueblo pots in the traditional manner. At the time of her study, each Pueblo had a characteristic style that determined the shapes of its pots and the colors and forms of decorations painted on them. (Bunzel studied pottery at the Pueblos of Zuni, Hopi, San Ildefonso, Acoma, Laguna, and possibly some others, with most of her time spent at Zuni.) The shapes of the pots of each Pueblo were made by coiling clay up from a base and thinning and smoothing the pot walls, a process different from the more mechanized approach of building pot shapes by throwing clay on a wheel. Clay for pot making was gathered from countryside near the Pueblos, as were minerals and metallic compounds for the pot colors. After a pot was shaped, it was covered with a colored clay slip and polished to a high sheen. Designs were painted on the surface, and the whole vessel was fired in an outdoor kiln.

Bunzel reports that although potters at the various Pueblos could not identify specific ratios that guided them in shaping their pots, they showed "a very definite sense of proportion" that resulted in a "uniformity of vessels from any one village" and a "distinctly critical attitude towards any deviation from the accepted forms": "From no woman did I get any rules of proportion. . . . The most general principles I could elicit were, 'It must be even all around, not larger on one side than another' (this from all potters without a single exception), the neck must not be too long, the mouth must not be too small" (p. 8).

Bunzel next sets forth basic principles of design that her research suggested guided Pueblo potters when they painted decorations on their pots. The task of painting, Bunzel tells us, is one that the potters spoke of as more difficult than building pot shapes: "Anyone can make a good shape, but you have to use your head in putting on the design" (a Pueblo potter) (p. 49).

Bunzel is interested primarily in the designs painted on the pots. She identifies, for each Pueblo, specific design limits within which the individual potters operate, beginning with Zuni:

> A Zuni artist sums up the general method of decorating a water jar as follows: "First I paint the stomach and then I paint the lips. I always use different designs on the lips and the stomach. You do not have to use the same number of designs on the lips as you use on the body." (p. 16)

In addition, at Zuni, Bunzel finds:

> It is quite clear in the minds of all artists that the number of designs is fixed, and that the patterns selected must be enlarged or reduced to a size suitable for the particular surface to be decorated. Along with this goes a very marked aversion to overcrowding. This is so very characteristic that I shall quote a few comments of informants:

I do not like to have the whole jar covered with paint. If I use large designs, I leave large spaces between, so that it won't look dirty.

There should be a good deal of white showing. If you put on too many small designs, the jar is too black and that is not nice. I do not use too much black because it makes the jar dirty looking.

If there is room, I sometimes use four big designs on a large jar. If the jar is very large, I use larger designs. I never use more than four . . . I like a lot of white showing. (p. 20)

At Acoma, the aesthetic is different. Bunzel comments:

We have already noted the tendency of Acoma designs to fill without break the whole of the decorative field, completely eliminating the background. This is the result of the way in which the Acoma painter approaches the problem of design. Instead of starting with certain familiar ornamental elements and constructing out of these a design that conforms to certain rules of composition, she reverses the process. Starting with the unbroken field as her decorative unit, she divides and redivides this into areas to receive paints of different colors. . . .
Whereas Zuni patterns are largely a handling of line, Acoma patterns are primarily a treatment of surfaces. . . . Wherever white occurs, it is part of the design. (pp. 35–36)

Design conventions are also suggested when potters from one Pueblo comment on the designs of potters at other Pueblos. Several Zuni women respond to pictures of pots from elsewhere that Bunzel shows them: "The deer look like rabbits." "The deer house is drawn wrong." "Someone did not know how to draw deer and put spirals there instead." "Deer are not good for the inside of a bowl." "Dirty looking." On the whole, however, Bunzels says, "the women found more to admire than to criticize" in other women's pots (p. 59).

Finally, after suggesting a backdrop of commonly accepted design principles that vary for each Pueblo, Bunzel turns to what she calls "the personal element" in design—the "frame of mind" in which an artist approaches decorating a pot: "We are now ready to return to our potter, whom we left some time back, holding in her hands the carefully molded and polished vessel, ready to receive its painted decoration. What will she do with this gleaming white or yellow or black surface? What is in her mind as she turns the vessel over in her hands, studying its proportions with reference to the style of decoration traditional in her group?" (p. 49).

Bunzel's discussion of the personal element in design is the heart of her book. Two dimensions of this personal element are most important to her. The first is source: where do Pueblo potters get their design ideas? The second is individuality: does a Pueblo potter see her pots as different from those of others? If so, how? A consideration of the individuality issue is fundamental to Bunzel's understanding of design sources. Bunzel notes that many of the pots from the same Pueblo look the same to outsiders and that her own attempts to identify pots by maker often failed. At Zuni, where she both studied the pottery making of others and learned how to make pots herself:

> I was quite unable to find any noticeable difference in style in the work of different individuals. The work of the professional potters showed a more perfect mastery of technique than the general run of Zuni pottery; the vessels were better formed, and more accurately painted, but they were not distinguished by individuality of style. (p. 65)

Bunzel looks at pots both in the field and in museum collections and concludes that the Zuni approach to design is "so formal that personal peculiarities of style would be displayed only in minute variation" (p. 65). To her, "the evidence of strong individualism is wanting in Zuni" (p. 66): individual differences are shown in a mastery of technique rather than in the invention of individually distinct design forms (p. 68).

When she next turns to another Pueblo, San Ildefonso,

Bunzel expects to find greater individualism because of the innovative black pottery work of Maria and Julian Martinez. However, similar problems appear. Although some San Ildefonso potters sign their work for the tourist trade, the signatures are frequently misleading:

> By no means all of the pottery that goes under Maria Martinez' name is hers. Much of it she has modelled; some of it is modelled by her two sisters; and practically all is painted by her husband Julian. (p. 66)

Of the pots of another San Ildefonso woman, Bunzel notes:

> [These] pots are, for the most part, decorated by [Tonita Cruz's] husband, Juan Cruz. Their work is almost as good as that of the Martinez family. Julian's designs are, on the whole, simpler than Juan's. Some, which are the last word in brevity of expression, are unmistakable. These excepted, it is not easy to tell the work of the two families apart. (p. 67)

Something thus seems to conspire to mock Bunzel's attempt to find familiar forms of individuality. One factor is that she is an outsider to the Pueblo pottery-making culture she studies. Perhaps signs of individuality known by insiders are not apparent to her. In addition, there may be other forces that cause Pueblo potters to deemphasize individuality in their work: a desire to protect collective ways, for instance.

Marriott's María: *The Signature Story*

A book on Maria Martinez by the anthropologist Alice Marriott[3] provides further detail that may help explain Bunzel's difficulty. Marriott describes an incident in 1909–1912 when a shopkeeper in Santa Fe asked Maria and Julian Martinez for more of their pottery to sell to summer tourists. Maria did not think they had more ready but suggested that her two sisters might. When the shopkeeper protested that buyers would want only Maria's work, Maria tells him, "It all looks alike to white people, my sisters' and mine. . . . You can sell it all just the same. You don't have to say who made it" (p. 200). The source of Maria's argument was her feeling that "It all comes from San Ildefonso. . . . It's all just alike in each pueblo. It's the pottery that makes the difference, not the woman who makes the pottery" (p. 200).

Fourteen years later, according to Marriott, Maria had a similar conversation with another white man:

> "Well," said the Superintendent, "do you know what I think you ought to do, Maria? When a white artist paints a picture or carves a statue, he signs his name to it. Then the people who buy it know it is his work. I think you ought to sign the pottery the same way. Could you do that?" (p. 233)

Maria again does not like the idea:

> "The pottery all comes from here [San Ildefonso]. It's all made the same way. It doesn't matter if it's mine or somebody else's."
>
> "Maybe not to you, and maybe not to the other ladies. . . . But people who buy away from here want to know." (p. 234)

Maria then agrees to sign her pots. Later, that night, she tells Julian:

> "Should the other ladies sign their pottery, too?" he asked.
>
> "I guess so," said Maria. "He didn't say anything about it."
>
> "It's all right," said Julian, "but suppose you sign your work and they sign theirs, and then theirs doesn't sell, what will they do? Won't their feelings be hurt?" (p. 235)

Maria answers:

> "White people can't tell much about pottery anyway. If the other ladies want me to sign their pottery, and they ask me to, I will. It's up to them. They can have it the way they like. I'll do what they want me to." (p. 235)

According to a note in Marriott, the date on which Maria Martinez first signed her pots is uncertain: "Jars signed 'Marie' were collected in 1923; unsigned jars date from as late as 1926. The 'Marie and Julian' signature first appeared about 1925" (p. 227).

Thus Marriott's account of Maria and her signature suggests both a desire to protect collective ways and an acknowledgment of internal differences. Maria does not say that her pots are indistinguishable from those of other women at her Pueblo. She says that white people will fail to distinguish them and that this, in her view, is desirable. Within each Pueblo, as Bunzel reports, individual potters acknowledge their differences:

> I can always tell by looking at a jar who made it.

> All the women use different designs.

> All the women paint differently.

> If I painted my bowls like everyone else, I might lose my bowl when I took dinner to the dancers in the plaza. I am the only person who makes a checkerboard design around the rim. (p. 65, Bunzel; specific Pueblo sources for each quote are not identified)

Commonality and Difference

In *The Pueblo Potter,* Bunzel tells of visiting an Indian trader who handled "the entire output of Hopi pottery" and could "name the maker of almost any one of the several hundred pieces in his storeroom." She was struck by the fact that

> Every day women bring in their work, and every day larger shipments are sent away. Women will trade a lot of a dozen or more pieces of pottery for credit at the store, and the pottery is generally sold in job lots. No stock record is kept, in fact, no record of any kind. It seems impossible that the trader should be able to tell the identity of each piece. (p. 65)

Yet the trader can tell the maker from the style of decoration, as can other potters:

> During the summer of 1925 I visited this store with one of the best potters in Hano in order to test her ability to distinguish individual work. . . . No attempt was made to identify the large number of small and badly made pieces, but she was able to tell the name of the makers of almost all the larger and more striking pieces. (p. 65)

Similarly, Bunzel is impressed when, at Laguna,

> One woman while looking over my photographs [of assorted older pots] stated that one of the bowls pictured had been made by her aunt. "My aunt always made designs like that." (p. 65)

Such recognition of individuality is striking, in Bunzel's view, because of the relative uniformity of design styles at each Pueblo and because similarities in style are valued over differences. Regarding sources of the individuality of designs that potters recognize, Bunzel notes that the potters speak of dreams. They also think of designs in odd moments while doing other things:[4]

> All [of the women] speak of sleepless nights spent thinking of designs for the pot to be decorated in the morning, of dreams of new patterns which on waking they try and often fail to recapture, and above all, the constant preoccupation with decorative problems even while they are engaged in other kinds of work. (p. 51)

The potters say:

> I am always thinking about designs, even when I am doing other things, and whenever I close my eyes, I see designs in front of me. I often dream of designs, and whenever I am ready to paint, I close my eyes and then the designs just come to me. I paint them as I see them. (Hopi)

I think about designs all the time. Sometimes when I have to paint a pot, I can't think what design to put on it. Then I go to bed thinking about it all the time. Then when I go to sleep I dream about designs. (Zuni)

While I am making a jar, I think all the time I am working with the clay about what kind of design I am going to paint on it. When I am ready to paint, I just sit and think what I shall paint. (Zuni) (p. 51)

Some of the potters report that to paint a design is to paint one's thoughts:

I get all my ideas from my thoughts. I think of my thoughts as a person who tells me what to do. I dream about designs too. Sometimes before I go to bed, I am thinking about how I shall paint the next piece, and then I dream about it. I remember the designs well enough to paint in the morning. That is why my designs are better than those of other women. Some people do not think that pottery is anything, but it means a great deal to me. It is something sacred. I try to paint all my thoughts on my pottery. (Laguna) (p. 51)

The most common source of learning about pottery making that the women refer to is one's mother:

I learned this design from my mother. I learned most of my designs from my mother. (Laguna) (p. 52)

Second come other relatives:

Nampeyo is my mother's sister and she teaches me designs. (Hopi) (p. 51)

I used to watch my aunt while she made pottery be-cause she was such a good potter. That is how I learned to paint. (Laguna) (p. 52)

Although they learn from others, the Pueblo potters do not approve of copying:

I make up all my designs and never copy. (Laguna)

I never use other women's designs and they never use mine. . . . I always have lots of designs in my head and never mix them. (Hopi)

I like best to make new designs. I never copy another woman's designs. (Zuni) (p. 52)

The potters also do not like to repeat their own designs:

My jars are all different. I don't make the same design twice. Sometimes I make two or three alike, but not often. I don't like to do that. (Acoma)

This is a new design. I learned the different parts of it from my mother, but they are put together in a new way. I always make new designs. I never copy the designs of other women. It is not right to do that. You must think out all your designs yourself. Only those who do not know copy. (Zuni) (p. 52)

Potters also use old pottery shards as sources for design ideas, and there too they speak of not simply copying:

I go down to the ancient village and pick up pieces of pottery and try to put them together and get the line of the design. (Hopi)

I save all the pieces of old pottery and try to work out the whole design from these scraps. Sometimes I use one of the old designs from around the rim of a jar and make the rest of the design out of my head. (Hopi)

When I find a piece of old pottery, I save it and get the design in my mind. (Hopi)

When I am ready to paint, I think how I am going to paint. I pick out pieces from the old village where I have my peach trees and I try to get the line of the design and think how it went. I put the pieces together and pick out the best. That is how I learned to paint, from copying the old designs. When I first started to paint, I always used the designs from the old pottery, but now I sometimes make up new designs of my own. (Hopi) (p. 52)[5]

For Bunzel, the important point in all this is that "each pot is approached as a new creation" (p. 52):

When a woman finds a piece of pottery with good painting, she tries to study out the "line of the design," her own imagination and her knowledge and feeling for the [old] style filling in the gaps in the very fragmentary material. (p. 55)

Concerning dreams, she notes:

When Hopi potters dream designs they are always in the accepted Sikyatki manner, just as twenty years ago they would have been in the style prevailing at that time. Nevertheless, from the standpoint of psychology of artistic creation, it is two entirely different things to dream a design and to copy one. (p. 54)

The Meanings of Designs

In addition to identifying design sources, Bunzel seeks meanings for specific designs. She finds these generally evasive. In Zuni, for example, "although it is possible to get a 'name' for any design," there is no consistent or "fixed terminology" (p. 53):

The same element is called in one composition "cloud," in another "flower," in yet another "drumstick." Furthermore, the same composition, with all its parts, will be differently named by the same person at different times. (p. 54)

At Acoma,

There is no trace whatever of symbolism in design. Even my most communicative informant could give no meanings of any kind. She said, "We have only three names for designs, red, black and striped." (p. 71)

At San Ildefonso,

> I had a series of designs identified by three different men, all of whom painted pottery. . . . To Juan all designs are clouds. Julian shows more imagination; the frequent occurrence of bird designations in his interpretations is notable. (p. 71)

Among the Hopi,

> The association between designs and objects or ideas is even more tenuous than at Zuni. Here, too, most women will find some significance in designs, when questioned, but the way in which the answer is given indicates that the association is quite secondary, and frequently suggested by the inquirer. (p. 70)

Other accounts, too, suggest a disconnection between pottery designs and any specific meaning attached to them:

> Although one can see symbolism in the [Pueblo] pottery designs, the names the Indians give to the patterns are primarily for non-Indians who ask for meanings.[6]

For Bunzel, the absence of a fixed terminology to describe the designs on the pots implies that decoration is "a sensual rather than an intellectual experience" (p. 54) and that it is present-oriented:

> There is no reason to assume that the meanings attached to Sikyatki designs [among the Hopi] are those originally associated with them, nor is any such claim made by the persons who use them. (p. 71)

Further, according to Bunzel, the potters' interpretations,

> for which they claim only a subjective reality, are strikingly different from those advanced by archaeologists. . . . The modern Hopi sees rainbows and

mountains where the archaeologist sees birds and ser-
pents. (p. 71)

Thus Bunzel seems intent on countering a view of fixed
meaning in design, and in substituting for it a suggestion that
the meaning of Pueblo pottery designs lies in the process of their
making. This process, in her view, is dependent, in significant
ways, on individual desires for self-expression. Bunzel discusses
designs more specifically than she does the construction of pot
shapes, although other accounts describe how these, too, get
individualized: "When I first came to know Maria, I would not
have known whose pots were whose; now there are certain ways
by which I can tell."[7]

Bunzel speaks of her experience of taking instruction in
pottery making at two of the Pueblos. At San Ildefonso, the
Pueblo women told her: "Go ahead and make"—"What shall I
make?"—"Anything you like" (p. 61). She is also told, "The
girls watch us make, and then when they want to make their
own, they make anything they want" (p. 61). She asks for help in
developing a design to paint on a pot and is told, "You must
think yourself what you want to paint. That is the way we do"
(p. 62). Bunzel finds herself reproducing, on a San Ildefonso–
style jar, a typical San Ildefonso paneled border: "Apparently in
even so short a period of observation and imitation (ten days), I
had assimilated the San Ildefonso style to such an extent that
I had unconsciously reproduced it when confronted by a vessel of
the familiar proportions" (p. 62).

Thus, for Bunzel, statements encouraging individual expression seem, at times, natural and, at other times, at odds with the
uniformity of the pottery that is produced at each Pueblo. Bunzel
seeks to settle the matter in various ways. At one point, she
concludes, "The impression of uniformity in the vessels of any
village is based on rather elusive traits. It is, in one sense, more
apparent than real. Gross measurements on a large number of
vessels show a considerable range of variation" (p. 8). The sense
here is that Bunzel is caught repeatedly in a struggle to reconcile
findings of individuality with evidence of standardization. Why
the need to do so? Is this Bunzel's emphasis or mine? What do
the potters think?

Reflections on the Individuality-Conformity Dilemma

The individuality-conformity dilemma, I suspect, is not the Pueblo potters' main concern, nor is it Bunzel's. In *The Pueblo Potter*, Bunzel is largely interested in identifying forms of individual expression in order to illustrate her contention that "primitive" art is much like that of more "sophisticated" societies: "Even among primitive peoples, art is recognized as primarily an individual function" (p. 1).

In reading Bunzel, I was struck with a sense that the individuality she points to is as hard to locate as the specific meanings of Pueblo pottery designs. Perhaps this is because Bunzel looks for individuality in a collective culture. Perhaps individuality is difficult to grasp in any context. Perhaps it is problematic for me in particular, independent of what anyone else thinks. When considering individuality in the Pueblo context, I keep wondering about how to separate the individual from the web of relationships that sustain her. I worry about the validity of doing so in a culture known to have other values. In my own context, I know I find it hard to separate myself—to experience myself in my own right. I am like the Indian potter painting her checkerboard design on a pot rim in order to tell her dish apart at village dances, yet I want to be like Georgia O'Keeffe: I want my difference not to be as small as a minor design element.

Bunzel, similarly, although she speaks of herself more formally than I do, must constantly reconcile her own sense of experience with that of the potters with whom she speaks:

> At First Mesa, I took one informant to the storeroom of the Hopi trader. . . . I asked my informant to select pieces whose purchase she recommended. Her selection presented the most motley assortment of good and bad painting, some of the pieces being not only unpleasing in design, but also slovenly in workmanship. I soon discovered that the selection was made entirely on the basis of the technical excellence of the ware. A handsome bowl by one of Nampeyo's daughters was discarded because one side had a slightly mottled appearance, showing that not all the water had been expelled in the

firing. "It will break when you use it." She admitted that the painting was good, but this seemed too unimportant to be commented upon until directly questioned. The prophecy was, alas, correct. I purchased the bowl for the sake of its design, and although I did get it to New York unbroken, it crumbled the first time it was used. She also advised strongly against the purchase of another bowl because it did not ring true, although when questioned, she admitted that this, too, was "pretty."

This emphasis on the technical rather than the artistic qualities of ceramic products is the common pueblo attitude. (p. 60)

Bunzel clearly tries to subordinate the words of Pueblo potters to her own more general description. That both her perspective and theirs emerge from her work occurs not only because Bunzel quotes from the potters' speech and seeks to capture their sensibility, but also because she reveals herself in varying ways: in the nature of her interpretations of others and in the personal details about herself that she lets fall. Bunzel is prejudiced: she uses words like "primitive," "sophisticated," "motley," and "slovenly." The Indians, for her, are often distanced others. As an anthropologist, Bunzel comes and goes and never feels entirely easy with the natives. The potters she studies are concerned with their art. Although Bunzel speaks less of her art than of theirs, similarities might be remarked upon. Bunzel, too, wishes to make a difference within a tradition. She wants to present an argument within anthropology concerning individuality and art. She wants to make others see what they might miss were she not to point it out. Bunzel is an author, an inscriber of her own perspective. *The Pueblo Potter* is her design, yet the originality of the account is constrained by the conventions of anthropological description and by the language its author uses to express herself. In her way, Bunzel dreams in the "accepted Sikyatki manner," all the while imagining that her dreams are her own.

If an individuality-conformity tension exists in Bunzel's relationship with her discipline, it also occurs in her relationship with the people she studies. In *The Pueblo Potter,* Bunzel tries to be respectful of her potters: to see things in terms of life as they know it. Thus she quotes from them and tries to describe their perspective. At the same time, she has her own ideas, and the boundary between the two is never entirely clear. Further, since none of this is talked about at length, readers have to guess at what is going on. What is Bunzel's perspective? What are the Pueblo potters' views? How do these make one? In figuring out such a complex expression, the reader then adds her own ideas: her own sense of the individuality-conformity dilemma, for instance, or her knowledge of Pueblo Indians, or of anthropology in the 1920s. This business of filling in the gaps surrounding a study is a way of making a study one's own, and it is as central to how social science gets done as are the efforts of any author to say exactly what she means. Finally, in Bunzel's case, as in that of her potters, the expression of individuality is never pure, and it is often a difficult achievement. It is probably impossible to break out of the clichés of one's time and place to assert a nonconforming individuality. Yet the attempt to do so may be an important aspiration because the clichés that surround each one of us never ultimately capture us, and we may wish for some degree of individual recognition, however small—for a sense that although we are "of" our surroundings, we must never be defined entirely by them.

Part II: Peterson's *Lucy M. Lewis*

Susan Peterson, a potter and biographer, has written a book about the Acoma potter Lucy Lewis.[8] Lewis, born about 1900, became broadly famous for her work in the 1960s. In the early 1980s, Peterson visited Lewis at Acoma. In 1984, she published *Lucy M. Lewis: American Indian Potter,* an account notable for its descriptions of the social settings of Pueblo pottery making. In her book, Peterson describes one of the kitchens where she finds Lewis at work:

> The kitchen table in Dolores' house is Lucy's studio. . . . [Lucy's] tools are simple, and she has had them for years: a few gourd tools—free-form shapes whose edges have been sharpened with a file; an old

sharp-edged metal can lid; some smooth pebbles; a
short knife; a leather tooling awl; a soft rag; a wooden
stick. [Lucy] keeps them all in a coffee can or a plastic
bowl. She works with a great economy of movement,
slowly and deliberately turning the clay, adding, sub-
tracting, pinching, and pushing with the experience of
seventy years. (p. 124)

According to Peterson, Lewis does not work alone:

Dolores [one of Lucy's daughters] has made clay especially for the piece Lucy is starting today. . . . [Lucy]
looks on quietly as Dolores rolls the first coil, long and
thin, across the table. . . .
    Lucy picks up the new coil, swirling it, squeezing
it gently so it will adhere to the one before. Dolores is
rolling faster now, to keep up. Emma [another of
Lucy's daughters] joins and the pace quickens. (p. 127)

Peterson describes many scenes of Lewis working in her own
kitchen, or in the kitchens of her two daughters, where Lewis
molds pots and decorates them with her daughters' help, although there are some aspects of the work that Lewis must do
herself: "Lucy works on the largest pot, a jar with bold black
zigzags. Dolores is filling in the deer Lucy has outlined. The
[younger women] do not know how to do fine-line decoration,
Lucy's specialty; that incredibly complicated geometric pattern
Lucy does all by herself" (p. 134). Thus individuality is important in Peterson's account. Peterson mentions that Lewis's daughters are known for their own individual styles of pot making:
"Emma is known for her small platters . . . Dolores is known for
her small canteens" (p. 128).
    However, more important to Peterson than the differences,
or similarities, between the pots of these women is the process
and setting of their pottery making:

Emma's own kitchen in her house below the mesa,
where she usually makes her pots, is a thoroughfare for
children playing, getting drinks, investigating the re-
frigerator, washing their hands in the sink, and look-

ing for things to do. When Lucy is there, some of them pull at Grandmother's sleeve. . . . From time to time Emma looks up from her potting. (p. 128)

Or,

> As the three women work at this kitchen table, six of Emma's grandchildren play nearby or pull at her, making requests. A new baby peers seriously from his cradleboard. Kittens surround their feet, wanting food. More children enter, home from school . . . [carrying] textbooks written in phonetic Keresan, part of the program to teach them their own language. The noise increases. (pp. 133–134)

In the end, Peterson reflects:

> How these ladies make pots says something of great importance to me, but it is something that did not become clear until I sat down and started writing. . . . As a studio potter and teacher, I have had much experience in potteries, workshops, studios, and classrooms. Such places are as familiar to me as my own skin. Yet, here are people who make pots in a place different from where such work is conventionally done. In every sense, these Indian ladies make pots in the midst of their lives. (pp. 134–135)

Elsewhere, Peterson describes another scene, with a different message. The setting is Dolores's kitchen:

> The conversation at the table has been going on for a long time. Dolores leaves and returns with a large plastic bag in her arms. Setting it down in front of her mother, she empties the contents. "Look what I bought yesterday from a shepherder, all these pot shards," she says, moving them around with her fingers. Lucy immediately picks up the fragments with fine-line painting or corrugated textures and begins to separate them, saying, "See how many different patterns they had? Those old people! See how nothing is the same, these are pieces

from different pots. How wonderful to have so much difference!" (p. 40)

The signature of the individual potter is thus hardly necessary for Lewis. The signature is in the pot, even when the pot is a joint product or found in pieces. I am reminded of Georgia O'Keeffe's rejection of signing her paintings. Although O'Keeffe's sense of self was highly individualistic, and Lewis's seems more socially embedded, each of these women recognized individuality when they saw it and each opposed a split between self and art. Peterson observes Lewis's involvement in her work: "Lucy works joyously. . . . When she is contemplating a new design, her concentration is deep; she may sit for hours looking at the shape of the pot in all its nuances. . . . The Acoma say that each pot has a spirit" (p. 139).

Peterson, in describing Lewis's art, repeatedly gives a feel for the potter's daily life:

> It has taken Lucy all morning to paint six fine-line rectangles, and there are perhaps fifty more to go. She sets the pot slowly on the table and turns it gently; she seems pleased with the results so far. Then she rises and withdraws to the back part of the house to rest for about thirty minutes. When she returns, she has changed into her finery: her sky-blue dress with the yellow orchid print, made of fabric brought to her from Hawaii; her crochet-trimmed apron; a turquoise manta pin on her breast; a large squash blossom necklace, earrings, bracelet, and rings—all of turquoise; a small red Hopi sash on her braid. Emma, who has dropped by for a short visit, stands up to comb her mother's hair.
>
> This is the last day I will be here to photograph her. . . . Lucy must have decided that she wanted to appear in print in her brightest things. (p. 138)

Why the description of clothes and finery? What does this have to do with pottery? When Peterson arrived to study Lewis, she had to deal with the fact that Lewis did not speak English, only Keresan. What Peterson learned about Lewis, she learned by listening to Lewis's children translate their mother's words, by hearing them speak about their mother, interviewing others, and observing Lewis at work. Dress is also a way of speaking, as is the decision to describe a daughter combing her mother's hair, or to mention a feeling that a person one studies cares about how she will appear in your book. Catching an image of someone you only slightly understand is a way of holding onto them. When I, like Peterson, do such a thing, I too write concretely: about what a person wore or how she looked. I forget, all too frequently, O'Keeffe's phrase: "the meaning of the flower *to me* at that particular time."[9] I forget that I am not primarily describing another person's reality, but my response to it, its meaning to me, and especially its emotional significance. It seems more important, at the moment, to feel that I am grasping someone else's meaning, or an external reality, even if that is not the case.

In part, I think, a desire to grasp an outer reality occurs because much of our training in social science is other-oriented. The external world always seems to count more than the inner world of any individual, particularly the individual conducting a study or writing a biography. The outside world is considered the real world, and it is looked to for security. It is as if, in the details of someone else's life, culture, setting, or story, or in a general description of a nation, an economy, or the world, the keys most centrally needed for the future will be found. In view of such big things and noble ideas, it is hard to acknowledge that Susan Peterson matters, and that, in the book *Lucy M. Lewis,* she matters more than the eighty-year-old potter she carefully describes. It is difficult to concede that Peterson's book tells us more about Peterson than it ever can about Lucy Lewis. Joan Didion has put it this way:

> We are brought up in the ethic that others, any others, all others, are by definition more interesting than ourselves; taught to be diffident, just this side of self-effacing. ("You're the least important person in the room and don't forget it," Jessica Mitford's governess would hiss in her ear on the advent of any social

occasion; I copied that into my notebook because it is only recently that I have been able to enter a room without hearing some such phrase in my inner ear.)[10]

I, too, was not taught very much about what to do when the directions change, when the observer becomes central to the scene she describes. Women especially are not taught this, since women are socialized to defer to others. However, social scientists, in general, are not taught how to deal with situations in which they are central, and one reason for this is that then the nature of our responsibility changes. The choices become increasingly clearly one's own. There is no more pretending that a study, or story, is about someone else, or that its basic imperatives are external. One has to admit to influencing one's work with an inner life that is often unknowable, or poorly known, but that is nonetheless critical to how one comprehends the world.

Part III: Trimble's Contemporary Potters

In attempting to place Lewis in context, Peterson refers to those other contemporary Pueblo potters who engage in the production of "sterile, standard, uniform ware" that is "fired in an electric kiln, where no smoke can 'mar' the pots" (p. 143). She summarizes the scene:

> Some Acoma potters are even buying greenware [pre-molded manufactured pots] from the hobby shops. Some paint with commercial ceramic colors rather than natural earth pigments. And some are using a clear glaze over the painting. "Most people don't know the difference," Emma contends. "Even at Indian Market we have seen prizes go to this kind of pot." (p. 143)

An older generation dealer says:

> I used to think you really had to be absolutely traditional in the firing. . . . But today I don't think the new dealers know the difference. The old dealers are in their seventies and eighties; we know how it was. Today the ladies of Acoma need money for their families—

more now than they used to. The dung-chip firing breaks pots, makes blemishes, causes smoke spots. Many dealers don't like that, and more important, they have trained the ladies not to like those things either. (p. 143)

In a book surveying potters' views in the 1980s,[11] Stephen Trimble, a journalist, picks up on various versions of the concern with maintaining traditional ways:

You've got to do it the right way. You can't get greedy. (p. 104)

The line break seems to be dying out because buyers think it makes the design unfinished. (p. 107)

The Pueblo women recognize commercial shortcuts because they make the pottery look "too perfect." (p. 107)

Juana Leno steadfastly follows the traditional Acoma ways. Her son operates a commercial ceramic supply at the pueblo. She put her foot down when he wanted to make a mold of her personal hallmark, the three-chambered canteen. (p. 106)

Trimble's book reflects a romantic vision of contemporary Pueblo potters and their work. According to Trimble:

Today, ancient traditions live on among the Pueblo people. . . . Pottery helps to bridge the gap between worlds, springing from old ways but generating an income in a wage-based society. Pueblo potters carry on with the grace, intuition, and eloquence of their ancestors. (p. 7)

The past, the present, and the future all are captured and united in the act of making pottery. (p. 12)

In one subsection, under the title "The Power of Individualists," however, Trimble quotes a Laguna potter speaking in a less traditional way of potters in her past:

> They were individuals just like me. One drew hearts, one drew something else. I can't believe it when they say that's traditional. I think it's just their inspiration—I wonder what this person thought about when they drew these things. I wonder what inspired this person to put these hearts on the pot. (pp. 81–82)

An Isleta potter has a similar orientation. Trimble comments:

> When Stella made the transition from commercial pots to traditional pots some twenty years ago, she developed her own style. Since she and her family are the most active traditional potters at Isleta, Isleta Pueblo pottery evolves in whatever direction Stella Teller's personal style develops. She now mixes her paints with the same white slip she uses to cover her pots. . . .
>
> "I am proud of my pueblo," she declares, "but I think of my pottery as Stella Teller pottery more than Isleta pottery, since it's my own style and the design ideas come from museum pieces, not from Isleta. Really, all the pueblos use the same motifs; they are individualized by the potters." (pp. 82–83)

Views such as these, emphasizing individuality, recur in the accounts of the contemporary Pueblo potters Trimble quotes, but far more common is an homage to tradition:[12]

> Al Qoyawayma goes to Hopi to gather his materials. He has a sense of history. "As I climb over the mesas and through the washes looking for clay, I realize that there have been many before me who have taken the same steps and made the same search." (p. 100)

> Potters pray before taking the clay; they make an offering of cornmeal, asking permission from Mother Earth. (p. 100)

Bessie Namoki holds out a smooth gray pebble: "A man from Second Mesa found this polishing stone at Awatori Ruin and gave it to my mother. Maybe in about 1937. . . . I don't know what I would do if I ever lose it." (p. 18)

Santana and Adam (Maria's oldest son), at seventy-five and eighty-three years old, still make pottery in the style of Maria and Julian. (p. 40)

Rose Gonzales taught her art to her son Tse-Pe and his wife Dora. First with Tse-Pe and now on her own, Dora Tse Pe Pena carries on Rose's traditions. (p. 43)

In the context of a desire to see the self as part of an art passed down through generations ("All the knowledge needs to be handed down because someday we'll be the old people" [p. 100]), the voice of individuality seems to emerge from out of nowhere: "From among the Sikyatki bowls coming out of excavation, Grace [Chapella] chose a butterfly pattern, added embellishments, and made it her own" (p. 95). Such a statement, like the act it describes, slips by very quickly. How exactly did Chapella make the pattern her own? Why did she chose the butterfly? Who noticed? What was the significance of her art?

In *The Pueblo Potter*, Bunzel's account suggested that identifying individuality in the Pueblo context might be difficult. Yet difficulty in grasping a sense of the individual is not peculiar to the Pueblo setting, nor is a feeling of relief upon finding it: "When Susanna Denet went to the Museum of Northern Arizona in Flagstaff, she was thrilled to be able to pick out her mother's pots from two rooms full of Hopi pottery. 'No one did the mouths like her. It made me cry to see those pots'" (Trimble, p. 95).

In social science, we also wish to recognize ourselves in our works, despite use of a generally impersonal style. We say that individuality matters. At the same time, we worry about whether our individual differences will be valued and whether what we say will be discounted precisely because of what we reveal about ourselves. We seek acceptance for our works and for ourselves as authors of them, but acceptance is only as good as what is offered for review: "When people accept my work, they are accepting a lot of me," says a Taos potter (Trimble, p. 34).

The question, I think, is, How can we give more in our studies than we have been accustomed to? How can the terms of social science be directed away from self-limitation? By giving more, I mean saying more about ourselves and about our individual visions, fantasies, conflicts, and attachments. I also mean speaking more in ways that are not clichéd, and that are not aimed primarily at conforming to standard theories and styles of speech. Social science is premised on minimizing the self, viewing it as a contaminant, transcending it, denying it, protecting its vulnerability, yet nonetheless mobilizing it as a tool for representing experience. To the extent that we constrain what we use of ourselves in our accounts, and to the extent that we effectively hide ourselves, we limit the ways we can depict others and we limit the entire social scientific effort. When we ridicule self-expression and call it narcissistic, unnecessary, or irrelevant, we box ourselves in, reduce the use of all our faculties, and make it less possible to achieve basic change in what we may know.

The Pueblo potters quoted here speak of pottery making, not of social science. However, I find in their comments a willingness to acknowledge the importance of the self in one's work in a way that is different from what is said in most discussions of the self in social science. As social scientists, we usually rationalize and apologize for ourselves. We paint pictures in which we hope not to exist; or, if we exist, our role is presented as subordinate, or as nearly invisible. The challenge, I think, is to change this posture so that our studies can become more honest. The works of Pueblo potters, and the accounts that describe them, do not tell us how to make a change from self-denial to self-acknowledgment, but they do suggest something about the nature of the conflicts and rewards that might be involved: "When you see the pot come out of the firing, it's unbelievable the high you get because you know that you've done it yourself. It's something that you just created out of nothing, just the clay, all made by hand" (a Taos potter) (Trimble, p. 35).

A Personal and Methodological Note

The accounts about, and by, Pueblo potters used in this chapter raise issues that concern me in my own research and writing. Like the potters, I, too, wish to make something I can take pride in and that may be recognized as mine. Like them, I often feel that words are inadequate as a form of self-expression. My desire is to speak without words, in a nonverbal form that nonetheless conveys my meaning. I would not say that writing for me is the same as shaping a pot is for a Pueblo potter, nor do I think it is exactly like painting a design on a pot. But there are some similarities. Most of all, I feel a kinship with the process the potters engage in. This process often proceeds largely by feel, although it has an intellectual dimension and is disciplined. I identify especially with the individuality-conformity dilemma that I see in the potters' situation, a dilemma that, in part, is suggested by the question, How can one feel like an individual when most of one's efforts conform?

In my social science writing and elsewhere, I know I try very hard to be different from others. At the same time, I am haunted constantly by a sense that, no matter how hard I try to assert my difference, most people, since they are outsiders to me, will think I am pretty much like everyone else. I probably compound this problem by being afraid to be too different. Like the Pueblo potters, I express myself through the conventions of highly stylized art forms: social science, for instance, or autobiographical writing.[13] My social science, although it is somewhat more candid than is the norm, stays on the safe side of a line, the other side of which would be to do work that describes my own experience, and that of others, in terms that would not be socially acceptable, or easily understood, terms that would not show concern for what others would think of me, or for whether I would be published or praised. I am always proving my worth in my writing, and investing a great deal of myself. Thus, like the potters in relation to their work, the line between maker and made is unclear for me. It does not really exist for me, except that I often think that my writing is better than I am.

Clearly, material from the Pueblo potters' situation enables me to articulate themes that have their source in my own life and work. The potters' accounts allow me to elaborate my own dilemmas and to see these as if they were outside myself and actually in the potters' experience. As I view the Pueblo potters, through accounts written about them, I identify with specific issues: the issue of signing a pot, for instance. In the process, I very easily forget that I used to wish not to sign my writing because then I would be saying that it mattered that I wrote it, and that I was buying into a system where names claimed works. What should matter, I thought, was that the writing was there, that it represented what it did, and not my attachment to it, or gains I sought to make because of it. However, simply because I forget such a thing does not make my underlying experience any less critical in affecting the descriptions that I write than if I was well aware of it. (My argument here is for self-acknowledgment, for owning one's influence on one's work, not necessarily for full self-awareness or full self-understanding.) Similarly, as I write of broader implications of the potters' experience for social science method, as I think, "What does a signature have to do with a sense of individuality?" I easily forget, or overlook, my own experience that is connected to that question: "What does my signature—my name, or what I call myself—have to do with who I am?"

This habit of blind self-articulation also affects other issues this chapter raises, the question of collective work, for instance: Where, or what, is individuality when a work is collectively made? There is the issue of sameness in design: Should a potter be seen as making up her own designs even if they look like everyone else's? The answer to this last question is of moment to me because it suggests that individuality is not to be found in appearances. Designs may look the same, says Bunzel, but "from the standpoint of [the] psychology of artistic creation, it is two entirely different things to dream a design and to copy one" (p. 54).

I find this statement of Bunzel's comforting. It speaks to my concern about the smallness of making only a minor design element that looks different from everyone else's. I know I am as likely as a Pueblo potter to say someone else's thoughts, thinking they are mine, to speak thoughts and feelings that are generally in the air (dreaming in "the accepted Sikyatki manner," for instance), and to think that my thoughts and dreams and feelings are mine and special to me. When someone tells me they are not mine and are common, I am hurt and I feel that I do not exist. When someone says, as Bunzel does in her quote, "It is two entirely different things to dream a design and to copy one," I feel I have been told that I exist and am different and special, and I feel relieved. I do not have to prove I exist by appearances. Thus, Bunzel's comment, because it relieves me, gets included in my account of the potters.

My personal issues, then, organize what I write, and this occurs in a largely unselfconscious fashion. My issues (the facets of my own psychological experience that are problematic or important to me) make me elaborate on some of the potters' experiences while ignoring others; they lead me to find great interest in small details and to want to present them. However, simply because my account is organized in a way that reflects my needs does not mean that it is invalid as a statement about something other than myself. Nor does it mean that the account—whatever it is about—has to be interpreted in exactly my way by others. Others may interpret my story in their own terms, miss my point, seek theirs, and make a sense that is very different from mine, even when speaking of these same Pueblo potters and drawing on my version of them to reach their own conclusions. Social science accounts, like the designs on Pueblo pots, do not have "fixed" meanings. "The same element is called in one composition 'cloud,' in another 'flower,' in yet another 'drumstick,' " says Bunzel, speaking of the designs on Pueblo pots. Furthermore, she adds, "the same composition . . . will be differently named by the same person at different times" (p. 54).

I am saying that such a design orientation is worth keeping in mind when we interpret studies.[14] I think we are used to a social science that is very much about who is right, or what lens, or view of the world, or of data, ought to be used. We are also accustomed to expecting that our studies will offer the kinds of simple arguments and propositions that make disputes easy to have. This is the hypothesis-testing tradition.[15] What is sought are assertions that can be argued with or supported, accepted or rejected, and potentially agreed upon. However, what if often the job is one of organizing evidence in order to suggest a whole that makes sense to a single author, and that basically takes care of its author? The author then is saying, "This, in general, is what something looks and feels like to me; this is what it is to me." The result is a picture that is more like a pot or a painting than it is like a hypothesis. It has themes or motifs, but it has no single or fixed point. My discussion of Pueblo potters is to be taken in this spirit.

10

Psychotherapy
and Pottery Making

Approaches to Self-Knowledge

WHEN I RETURNED from my last trip to New Mexico, I brought
into one of my therapy sessions a book showing pictures of Pueblo
Indian pottery. My therapist especially liked the older pots dating
from nearly a thousand years ago. I told her about how the
pottery was made—the traditional way of shaping, painting, and
firing it—and I told her that Bunzel's book had quotes from
potters about their dreams. My therapist was particularly interested in the dreams and asked for a copy of the pages describing
them. I said that in the 1920s, an anthropologist would ask
artists about their dreams, while in the 1980s, we do not normally do this.

In sharing my pictures of the Pueblo pots with my therapist,
showing her a pot I brought back for myself and one I wanted to
give her, I was telling her about myself. She was telling me about
herself, too, in being interested. In the back of my mind, I feared
she was thinking in terms of objects, egos, meanings, that she was
seeing me psychoanalytically in relation to the books and pots and
her. I wished that, instead, she knew more simply what my sharing
my books and my pots said about my relationship with her just
then: that I was cautious still in letting her know what was special
to me. She looked closely at my pictures and the pots, and I felt
that, for the moment, she was appreciating the artifacts more than
anything else. She once studied music, respects art and those who

make it, and likes old-fashioned things. I wish to say that the setting counts, as do the details: pots, books, dreams, therapist, feelings.

I wish to say, too, that speaking in a language of concrete images is a way to make a social scientific statement, and that it is a more comfortable way than many others for me; it is preferable to speaking in more general theoretical terms, for instance. In discussing my experiences of psychotherapy, I am also speaking of issues that can arise elsewhere, but always when these issues arise, they appear in particular settings. It is the nature of the description of any instance that counts (what the description calls attention to, what it says, or means), not who the description is about (an author, an "other," a potter, a therapist). It is not the level of abstraction a description uses that counts most either. A good description can be general or particular, and always is implicitly both: the particular description has systematic biases that give it a general shape; the general description has specific roots. In describing psychotherapy experiences I have had, I wish to emphasize that the self is known in relation to others and according to specific sets of ideas.

My therapist, to whom I feel indebted far more than I can say, was trained as a psychiatric social worker. She is more fond of abstract ideas than I am, and she knows the older depth psychology theories better than the newer ones. Recently, she suggested that I read Kernberg (on splitting) and Mahler (on separation). She also referred me to Winnicott on the role of transitional objects in creativity. On my trip to New Mexico, I stayed with a friend who is also a therapist and who had, on her shelves, some of the recommended books.[1] I had read them before, but I now reread them. I still did not like Kernberg; I was tired of Mahler. I had arguments with all their arguments. Back home, when I finally obtained the right volume of Winnicott from the university library,[2] I was not happy with that either, but I did not tell my therapist about that for some time. I did not want a difference of opinion to come between us. It did not occur to me, just then, that finding other people's theories lacking might be a sign primarily that these theories did not speak to my need for my own way of thinking, rather than a reflection of grave deficiencies in the theories, or in the person who recommended them.

I now read Kohut more extensively than I had before.[3] A previous therapist I had seen had liked him, and I knew he spoke of the self. I could recognize an affinity between Kohut's concerns and feminist thought because Kohut's theories are wholistic (the person's self is a whole) and Kohut emphasized a connection between self and other, as did Gilligan in *In a Different Voice,* for instance.[4] Nonetheless, I had a preference for the clarity of Gilligan over the psychoanalytic wars Kohut seemed engaged in. Gilligan, like other feminist scholars,[5] criticizes modern concepts of the individual because they hold up as normal a goal of individual autonomy, rather than acknowledging the webs of social connection that make individual personhood possible.

Gilligan's point is that individual autonomy models do not fit the lives of many women; they are based on male experiences. My sense is that these models do not fit anyone well. In thinking about individuality, I would rather not ask whether a person has become separate from others, but, instead, what the nature of the person's relationships are. Knowing oneself, and having a sense of self, then becomes less of a lonely business, and it feels more realistic to me. Self-knowledge changes from being a matter of discovering who, apart from others, one is, to a matter of recognizing how one is in relation to others and the external world. This leads to the identification of needs and to learning about the nature of one's dependencies. It can also lead to a recognition of the importance of having inner aspirations.

A Pueblo potter, for instance, says of her clay: "The clay is very selfish. It will form itself to what the clay wants to be" (Santa Clara). She might be speaking of herself. Another potter says, "A woman can make anything, any kind of shape with her own hands" (Santa Clara). A potter thinks: "My father has a big water jug that my grandmother carried from Oraibi to Hotevilla when they left. It sits in the bookcase by my father's bed. Sometimes I just look at it and think—someday I'll do something like that" (Hopi). Potters refer to "the feeling of creating, the part of life that makes you happy and joyful. . . . I feel the same way when I look at my kids or make a beautiful pie" (Santa Clara).[6]

Psychotherapy, like pottery making, is a way of creating and of expressing inner experience. However, it is a form of expression that emphasizes an intellectual grasp of the self. It is an applied social science (clinical psychology), and it shares its dominant values with social science in general. In theoretical terms, psychotherapy, of the type based in analytic psychology, is perhaps the Western world's most well-articulated attempt to understand the self. On the surface, such psychotherapy is an almost opposite approach to that of the Pueblo potter who comes to a sense of herself through the nonverbal work of pot making or design. The person involved in psychotherapy comes to self-knowledge through words and with reference to specific theories about the self. A therapist typically uses theories, often explicitly and self-consciously, to guide the understandings she and a client achieve. Nonetheless, the potter's nonverbal, and nontheoretical, approach may come into play in psychotherapy as well.

I speak of psychotherapy both because it has been an important part of my social science education and because it is an effort to arrive at explicit understandings of the self. In psychotherapy, the authorial self is central. I am referring to the self of the client who is seeking understanding. In much of the rest of social science, the authorial self is secondary. Knowledge about this self is viewed as a means to the end of understanding others, and not as an end in itself. The psychotherapeutic process—devoted to developing explicit understandings of the client's self—seems to me interesting both because of its subject (the participating, or authorial, self) and because of its style. Its style contains elements of a traditional social scientific approach and of a less traditional approach. The traditional social science approach is intellectual, and it aims at generally acceptable, and sharable, knowledge. Less traditional for social science is an approach that is emotional and that aims at individual knowledge. Psychotherapy has a traditional aspect that may be likened to the stance of a traditional academic. The less traditional aspect of the process, to my mind, has affinities with the approach of a Pueblo potter.

A psychotherapy process can be seen as like a pottery-making process in the sense that it is guided by feelings that often cannot be comprehended by explicit theoretical statements. These feelings give the psychotherapy process a shape and make it, to a large extent, an emotional product. It is something that is felt more accurately than it is verbally, or theoretically, known.

The verbal and theoretical statements made in psychotherapy grasp only small pieces of the client's sense of self, or of the therapy process, and they grasp these experiences only fleetingly. Further, the accomplishments of psychotherapy often occur in emotional moments rather than in intellectual ones. These are moments of inner recognition, rather than of expressed insight. In these inner moments, the client recognizes that something is being made, or known, and that this product is related to her sense of self. The psychotherapy process, like a pot one is part of, carries its maker's thoughts, feelings, and spirit and brings these to a new resolution.

Second, the psychotherapy process, like the Pueblo pottery-making process, is the product of an ongoing interpersonal relationship. The knowledge it generates is critically affected by the relationship between a therapist and client, much as a Pueblo pot is affected by interpersonal relationships in its collaborative setting. Third, the self-knowledge gained as a result of the joint effort of therapist and client is not by-the-books knowledge, but practical knowledge. Like the knowledge gained in pottery making, psychotherapeutic knowledge is guided by a practical sense of "how things ought to be" that is often different from a formal sense, or from what is prescribed by a formal theory. Bunzel notes, for instance, that the Zuni potters she studied, in talking of their painting, had "very decided number preferences" concerning the number of designs to be used on specific portions of a pot, often expressing a preference for the number four.[7] However, as she examined pots, she found that the actual numerical preferences were not the ones expressed: "In simple arrangements, anything from three to nine may be used."[8]

Similarly, a theory in psychotherapy may call for client and therapist to follow a strict protocol: to keep clear the boundary between self and other, for example. However, in practice, both therapist and client may benefit from an inability to keep boundary distinctions. In part, this is a statement of a difference between actual and idealized practice. But it is also a statement about two layers in what is going on. One is the layer that can be abstractly described and put into words (the strict protocol, the four designs). The second is the layer that escapes the formal wording, but that is nonetheless a crucial part of the reality: the experience of muddled boundaries, or the use of three to nine designs.

In Pueblo pottery making, the outcome most remarked upon is a pot with a finish and design. In psychotherapy, it is an improvement in a client's well-being. However, in both cases, a recognition of self is central to the process, and this recognition may also be viewed as an outcome. Both psychotherapy and pottery-making processes rely on a sense of seeing oneself in what one has made: in a pot design or in a psychotherapeutic insight, for instance. The design and the insight each provide a momentary view of the self through something seemingly outside the self. Each seems to be one's own, at least in part. Each is clearly a product that involves the self, and one knows about this involvement because one feels it, not because it is intellectually so, or because of the words one uses to describe it.

Finally, in psychotherapy, as in pottery making, the artist, or author, seeks not to copy other women's designs. The psychotherapy client, like the potter, seeks designs that she can call her own. A theory about the self used in psychotherapy may apply to other people, too, for instance, but the person involved in psychotherapy wants to feel that what she does with the theory is specifically her own. Bunzel says of a potter's design sketches: "These are not design elements in any sense that would satisfy a sophisticated analyst of design. They are patterns adapted to use on special parts of vessels."[9] Something similar might be said of psychotherapeutic insights. In psychotherapy, one seeks forms of knowledge that are special to the self, and that the self can recognize as its own. One seeks to avoid repeating insights because the self in psychotherapy is in a process of change. The self uses the psychotherapy process, like a potter uses a pottery-making process, as a means to fuel, or keep up with, change. These are some of the ways psychotherapy is like pottery making: it is emotional beyond words, collaborative, practical, self-recognizing, and self-involving, and it is, to some extent, original and involved in change. These are also ways that social science, more broadly, is like pottery making. Social science, too, contains elements of the Pueblo potter's approach and of the traditional academic's approach. It is emotional in content and about private knowledge, at the same time as it is intellectual and about knowledge that can be shared.

My general comments about social science, pottery making, and psychotherapy are based on specific experiences. I have been engaged in psychotherapy for some time, and my understanding of that process has affected my social science, not only initially, as part of my basic education, but on a continuing basis and as something seen in retrospect. In the psychotherapy I have engaged in, I think I have had more of a potter's orientation than a traditional academic's. From the start, I seemed not to come away from therapy sessions with clear self-understandings. I gained some insights, but what mattered to me most was my relationship with my therapist, and that relationship was something I sensed, and felt, far better than I intellectually knew it. I was moved when my first therapist viewed me with affection and interpreted me kindly, for instance. I tried to reproduce her attitude in viewing myself and generally, diffusely, took the therapy in. It sometimes troubled me that other people I knew who saw therapists seemed to know what they were learning. They could point to major insights, or to gains in personal development, when mostly what I could say was that I was glad my appointment was regular and that my therapist acted warmly toward me.

Later on, while seeing another therapist when I was teaching in the Southwest, I was more mechanistic in my approach to dealing with the psychotherapeutic process. I identified distinct behavioral patterns that I thought caused me difficulty, and I tried to change them. In retrospect, I think I gave more weight to the usefulness of such self-improvement engineering than time has seemed to warrant. I thought that I could change my inner psychological problems by manipulating external behaviors, both my own and those of others. More significant, however, than the particular strategy of change was my understanding of the source of many of my problems. The source, I felt, was a lack of separation between self and other. I felt I had to determine which emotions were someone else's and which were mine in relationships and then draw a line between them, as if this would solve things, or at least lessen confusion, the second of which it probably did.

Today, however, I find it less urgent to feel separate from others than to feel connected, not because I have learned that neat separations between people are impossible (although I think they are), but because I find that the quest for separateness tends to deny connection and to result in a sense of isolation for me. When isolated, I easily become scared, depressed, and paralyzed.

It is difficult to describe a therapeutic process lasting fourteen years and involving four therapists, two of whom I left behind when I moved about for jobs, and one who died of cancer. If I were a potter, I might make a pot to show something of my feelings toward these therapists and toward this central experience of my life. It is an experience that affects my social science not, for the most part, because it is about the self, but because it is about ways of knowing: about fairness, and process, and testing theory against evidence, and especially about considering evidence one does not feel detached from.

The self-knowledge I have gained from psychotherapy is something that has come about in bits and pieces and not very intellectually. As soon as I think I have gained an idea about why I have acted or felt in a particular way, that idea is gone. Another thought, or a different sense of experience, takes its place. This psychotherapeutic self-knowledge is like the design on a Pueblo pot: it is "pretty" but only part of the story. The larger story is the background experience that is too complex to be comprehended by any one description, and that constantly changes. Both its form changes (the nature of one's experience changes) and its need changes (the interpretations I may find useful at one time seem less useful later and I need new ones). Psychotherapeutic self-knowledge is also, for me, very much like a pot that breaks in firing. Its nature is suggested in comments made by Pueblo potters about how they feel when they lose their pots to the fire:

> Pueblo people build a kiln from scratch each time they ground-fire their pots.

> Firing days are dry and calm; a hot, clean fire is critical.

> Berenice Suazo-Naranjo says . . . "Sometimes you have a special feeling for a pot and you hate to fire it because you are so afraid it will break."

Everyone hates to hear the sound of a pot "popping"—going off like a "bomb" [in the fire].

Emma Mitchell doesn't "feel bad about the pots we lose in firing because we recycle the ones that blew up in the fire. We use it as temper and it just goes right back into the clay."

Nearly every potter has a few cracked pots in her home, pieces she cannot sell for a good price. All feel great affection for them.

Candelaria Gauchupin told her daughter that pots "are like people. We people have imperfections. Do we destroy them because they are blind or deaf or have lost a limb? No, we don't destroy them. She told me that pots are the same way. You don't destroy a pot because it has an imperfection. You love it as much as you would a perfect pot."[10]

Self-knowledge, then, like a pot that breaks in firing, and like any other type of knowledge, is not a fixed entity. It is there one minute and disappears the next, to be mixed back in with other clay to form a new pot. Or it is kept, despite its imperfections, held, like the self it expresses, and viewed with affection.

Self-knowledge is also like writing traditionally was for the Pueblos. As Peterson comments:

I have often encountered people from various pueblos who distrust attempts to put their own languages into writing for fear of the mistakes and sacrifices implicit in the written word. . . . Writing can record words, certainly, but cannot fully transmit an oral tradition. Voice, emphasis, tone, rhythm, facial expression, gesture, atmosphere, and many other things all convey meaning and nuance. Constant and continuous personal contact, rather than the written word, is perhaps one factor at the heart of the identity and longevity of the Pueblo people.[11]

I am suggesting here that in pottery making, psychotherapy, and social science, words can fail to grasp meaning, that
the self of a knower matters more than what is known, and that
this self exists in relationships. For example, Georgia O'Keeffe
did not abandon civilization when she left the East for New
Mexico. She did, however, experience different sorts of connections after she moved away. The Southwest suited her, she
painted it, and she insisted that what drove her was both her
work and her sense of self:

Taos, Summer 1929

Dear Henry McBride:

[I'm] finally feeling in the right place again—I feel like
myself—and I like it. [12]

Feeling like oneself thus depends on context. When O'Keeffe
left New York for the Southwest, the place where she chose to
resettle was near several Pueblos where Indian women made pottery. Yet Pueblo potters and Georgia O'Keeffe are not often
thought of in a similar fashion. When we think of O'Keeffe, we
tend to assume she is significant because she broke with tradition
and created an art that was shockingly her own. We view Pueblo
potters as important because of the tradition they carry on.
O'Keeffe is seen as an individual, almost without a culture, creating her own world in isolation from others. Pueblo potters are seen
as fighting to maintain their world, collectively attempting to
keep their culture from getting lost. However, such a contrast
between the two is perhaps too stark. It might be more instructive
to consider the relationships that the potter and the painter each
need for the creation of their art. Some of those relationships are
external. They are with other people who support one's work or
art, or with a countryside or a tradition (abstract painting for
O'Keeffe, pottery making for the Pueblos). Other sustaining relationships, however, are internal and are created in a space that
others either refuse to, or cannot, occupy, an inner space that is the
artist's own experience. O'Keeffe speaks of looking for someone
outside herself to tell her how to paint a landscape:

I thought someone could tell me how to paint a landscape, but I never found that person. I had to just
settle down and try. I thought someone could tell me
how, but I found nobody could. They could tell you
how they painted their landscape, but they couldn't
tell me to paint mine.[13]

A Pueblo potter says:

If I were teaching a young girl, I should tell her that
she must be most careful with the polishing. In painting I should tell her to use her own brain and to paint
any kind of design she likes. (San Ildefonso)[14]

The potter thus tells the student that there is an area in which she
is expected to make something of her own, a design whose dimensions cannot be prescribed for her by anyone else. Nonetheless,
someone else can offer a relationship that encourages the student's
individual creative activity, in part by identifying that activity as
the student's own.

When copying is not the point, when internal landscapes
are looked to for external form, the crucial matter is not whether
the individual is autonomous or fully known, but whether she
can learn to draw from herself. O'Keeffe says:

And then I wonder what painting is all about What
will I do with those bones and sticks and stones—and
the big pink sea shell that I got from an indian—it
looks like a rose. . . .

When I leave the landscape it seems I am going to
work with these things that I now think feel so much
like it.[15]

A Pueblo potter says:

I had represented the sun and the moon from both
sides and then I looked at the sky, and to me the sky
is smooth. Then looking at the ground—the sand,
the rough places, the feeling of walking without
shoes. . . . So I molded the pot, trying to make that
pot look more like the world. . . . If they just feel it,
maybe they'll get closer to what I mean. (Hopi-Tewa)[16]

The potter and the painter each mold the world, looking at the ground, borrowing bones and stones and sky. Each seeks to make a shape, and then someone else can get closer to what the artist means. It is not the outer world we see rendered, then, but that world filtered through what a potter can know. The question for social science is, What will we let our artists know? What will we let them express? Will we draw a line around our personal experiences and continue to stigmatize statements about them, or will we do better? Many processes may contribute to an individual's sense of self, worth, and perspective on environment: psychotherapy, pottery making, writing, research. If social science reflects its makers' worlds more than any others, it is perhaps time to acknowledge in detail, and with more honesty than we have before, the ways that our outer depictions reflect the nature of our inner lives.

IV

Teaching and Research

Experiences in Teaching

Exposure, Invisibility, and Writing Personally

The Professor

IN THE BACK OF MY MIND is a college professor in a long, gray overcoat. I found him as an undergraduate and he has not yet left me. He, and he is a he, walks into a classroom and mumbles a few words into the collar of his coat and in the general direction of waiting students. Whether the students can make out all his words is unimportant. It is their responsibility to fill in what they have missed or to go with what they have heard. The professor does not stay long in the classroom. Sometimes he says only one sentence. The students do not come late, and they are expected not to be absent. The professor's words may be obtained from someone else but, in my fantasy, they never are. Seeing the professor is a necessary part of the experience.

The professor's sentence (it is occasionally a single word) is sufficient to prompt reading, thinking, writing, and learning on the part of the students for the entire subsequent week. Or to prompt it for the next two weeks if the professor, for some reason (perhaps he is tired), cannot make it to the next class session. For pedagogical purposes, the professor occasionally chooses not to come to class at all.

That is the story, or the image. I have recently tried to put a woman's face, usually my own, into the overcoat, proving that a woman, as well as a man, could be the professor. When I do this, I am thinking of choices I would like: I want to be able to be this man.

I imagine that my professor wears an overcoat because he is teaching at a campus in New England. It is winter and snowing outside. He, or the room that he enters and stays in briefly, is cold. Yet his clothing is not only a result of the temperature. The professor wishes to hide and finds the protection of his overcoat comforting. He is largely unaware that his coat keeps others from knowing him, or from thinking that they know him because they see him. What matters to him is nothing personal, but, rather, what the students make of what he puts before them.

I never had a professor who looked, or acted, exactly like this overcoated man. Thus, for a long time, I thought he was me—that he represented my wish to teach without revealing myself and without having to deal with the psychologies of other people. That I did have professors and schoolteachers like this man has been an afterthought. These were people who did not wear outdoor clothes in class and who spoke more words than my professor, but who acted similarly underneath: they kept their inner selves hidden. My overcoated male image crystallizes them into one. This image, I still think, is what being a professor ought to mean.

I have now been teaching classes on and off for thirteen years, part time and temporarily, in various parts of the country. I have never had a full-time teaching appointment that lasted for more than nine months. Thus each time I teach I start fresh, making things up as I go along, rather than following a routine. I think of teaching as social work, like conducting a community meeting. My model for how to teach comes from psychotherapy, where everything is one on one and process matters more than professing. I structure the learning process of a class with research assignments, telling the students that what counts most is not what I say, but what they conclude from their own experiences.

Teaching is not really me, I feel. I do not like center stage. I am afraid of people, especially in groups. I forget that the students see me in the classroom, and when I remember, I find it hard to complete my sentences while speaking. I am reluctant to state my opinion. I feel that a class should be left in its pure form, as if I were not there.

I ask students to speak of their feelings in class: "How does this strike you? How did you feel when you read it? What is the importance of this research project to you? Write your papers in the first person." I pose for them the same basic questions I ask myself, or that I ought to be asking. I also treat students like I treat the people I research: I listen to their answers, take notes, observe. I want to be let into their world, although it frightens me because it is unfamiliar and I feel left out of it in the end.

Recently I have found that I have to speak of myself in my classes. However, personal revelation is not comfortable for me. I fear the spotlight. I fear rejection. I feel that my emotions, and opinions, ought not to be brought before a class. I feel that to use emotions to communicate is to pull a trick. It is manipulative: something women do.

Yet despite these feelings, I have now had enough experience teaching to suggest that I do best when I deal with things personally, even if this is difficult for me. Why, then, do I still carry an overcoated teacher into every classroom I enter and regret that I cannot be like him? Why do I feel that the social part of teaching is not the point, that the attention to emotions and anxieties, the need for reassurance on the part of both students and myself, ought not be there? Why is it so hard to learn that the overcoat, as a form of protection, fits another person, a man, and not me?

Exposure

It is fall twelve years ago and I am teaching for one semester at a university in the Midwest. I have been hired to offer an undergraduate course on popular culture and a graduate course on qualitative method. In the undergraduate course, a student takes pictures midway through the semester and, seeing them, I am amazed that the class could take me seriously. The photos show someone with kinky hair long as it would grow, sticking straight up and out. I think maybe I get by in spite of such a look because of my seriousness and my conviction that I, at least, think I am doing something right. I send the students out to observe and to read so they will be able to talk in class instead of me talking. There are forty of them. They sit in rows of bolted-down chairs facing front. I wish I were not standing before them, and I try different tactics to make them take my place in the front of the classroom and speak to each other.

In the doctoral seminar, which is smaller (eight students) and meets around a table, I keep giving the students exercises: "Here are four pieces of evidence. Do some research and invent a story in which they make sense." When we share the stories, no two are alike. In that class, one student becomes upset when I give an assignment that involves self-reflection. I have asked each student to consider a taken-for-granted understanding they have of themselves and to speculate on how their research or writing might be different were that understanding to change. The student who objects to this exercise calls the assignment, and later me, unethical. I am startled but gather what she means: classrooms and courses are not places where students should be expected to talk about themselves. These are public settings, and a student's inner life is something to be kept private.

To try to settle the matter raised by this student, I spent portions of several subsequent class sessions on the issue of personal examination. The other students in the class had not objected to the self-reflection assignment, but they humored me and the dissenting student in discussing it. The woman who was upset later filed a grievance with a university committee concerning my ethics. I wrote to myself in a diarylike story I was working on at the time:

> In my mailbox this evening was a letter from a student wanting to know her grade. She is one who has given me trouble over an assignment asking the graduate class to examine an understanding of themselves. . . . I read her letter and came back to the main office where she was working and sat down with her and talked. She was calmer than last week, as she had been even in class this morning. We may have talked again for two hours, but more slowly this time, and I could say I often felt as she did, that it was my right not to have to

give pieces of myself when asked, that I, too, treated classes with difficulty, becoming a more fearful person in them. But I was coming from a place where I felt insularity hurt me and she was coming from a place where exposure hurt more.[1]

In the stories I wrote reflecting on my teaching experiences, it was clear that I was having to confront a difference between how I might teach and what others expected of me:

One of the other students in the graduate class, who has been teaching for five years, tells me he has his objectives and expectations for teaching down pretty clearly by now and suggests I do the same. I try to get my objectives clearer, try to make a list like his, but feel it will never function as well. Yesterday morning in class, I read my list.

One student said afterward maybe I had objectives for open discussion, but it turned out I had ideas about how talk should go and he could see conflict there. Someone else said it might help if I gave my comments on the readings at the start of the Thursday sessions where we discuss them each week, rather than waiting until the rest have given theirs. It was a surprisingly practical response. Three said my desire to allow some nonjudgmental space would not work: this is a university, judgments are central to it, that is not a kind of space they can have. Yet two say they already have it in the class. We stay over the hour talking not of research but of education, and not feeling expert enough, I leave.[2]

The notes show that I wanted to get beneath the roles the students played:

After class, I ask Jim why he plays student so obviously with me, asking often what I expect, and if I know enough to tell him, trying to do just that. He says he is very aware of the roles because of having to switch from one to the other, and that he has to be to make it work.

I am impressed with the usefulness of that tactic but also have a sense that the "student" Jim is a model of the student he sets himself up to have, and it is the Jim who makes up the model I must know.[3]

The students were probing me as well:

Last night the class met in an extra session to make up for one we had missed. It was different from our other classes because after two people made presentations, and two had left, after someone had put the apples and cheese and what was left of the beer in the center of the table, there developed a kind of talk I normally would not engage in, because it ran late and was intense and, in some ways I care about, undisciplined, and because it put me in the position of defending my style. At one point, the woman who has trouble with me said "that is not social science," and she seemed puzzled more than angry. I said I thought it was, making a plea for acceptance in the only way I usually try, as an individual statement. "This is what I do," I say, which is a poor defense when what is wanted is a statement of fit to kind, so that others may know where I stand among the choices, and perhaps more importantly, where among the chances for making it. But I know this too little myself.[4]

In this particular night session, several of the students had wanted me to define something they called my "framework." They wanted to know how I thought: about social science, society. What was the structure of my thinking? I felt they were looking for a system. I could not answer and I left the session early.

I stayed on at that midwestern university for a second semester because somebody knew somebody in another department and I could teach a basic organizational behavior course. The department that had hired me for the fall, primarily for the qualitative methods course, had, at the same time, also hired a man to teach quantitative methods. He was kept on. I was not. I howled loudly, crying into the rain as I walked home the night I heard the final verdict. I had not wanted to stay in the Midwest, but I did not want to be kicked out either. I got a haircut. I told myself that the haircut was because of the harsh midwestern winds, which tore up my hair and would not let it lie flat. In the spring, while teaching two organizational behavior courses, I revised my dissertation. I locked the door to my office, avoided students, and worked on cutting the dissertation down to book length. In my mind, I did not value teaching, or myself as a teacher. I valued writing.

Fitting In

The next fall I was in the Southwest. I was at my first teaching job that was to last for an entire year. In preparing for my courses, I inquired about textbooks on interviewing and organizational communication, the subjects I was to teach, and then ordered the recommended books from the campus bookstore. As courses began, the books became a problem because I had trouble reading them. I came to my classes with a vague hope that the students had read what I had not. My reluctance to read was especially troublesome in the organizational communication course because it was a graduate seminar and relied on reading. The interviewing course, two sections of it, relied more on research. I told myself that what I read did not matter. The textbooks were for the students: they were the ones who wanted an overall, rational, impersonal view with diagrams and charts. I wanted mainly to hold onto a job and to make life easier for myself, which textbooks were supposed to do. That was why I had chosen them.

One day about six weeks into the fall semester, in the organizational communication course, one of the students came up to me. He was a military man who worked at a local air force base and came to class in his uniform. For several weeks, he had been challenging me with pesty questions from the back of the room. He was expressing his dissatisfaction, and I usually tried to ignore him. Now, during a break halfway through the class session, he approached me in the hallway outside our classroom. "Why don't you teach us what you are most expert in?" he said. "It can be anything. It doesn't have to be on the subject. If you teach us what you know best, we will learn the most from you. You see, what we want to do is to learn from you."

I was shocked. However, I took his comment seriously. I chucked the textbook, dug up some articles I liked, and put them on reserve for the students. The class dynamic changed: the students were more relaxed and possibly I was. We talked about how we felt about what we read and why it interested each of us. The lesson was clear to me at the time, and I have recalled it often since then in attempting to guide myself on how to teach. This student was saying, "Don't try to be conventional. Don't do what you think you ought to do. Be yourself." Nonetheless, I still find that advice hard to follow. In subsequent years, on several occasions, I have used textbooks, and at no point have I been able to read them carefully. But the issue is about more than reading. That student in his military uniform was challenging my whole effort to fit in, my attempt to wear an academic uniform. He was telling me not to try to pass: that I ought not attempt to do things in a standard way if only because I could not pull it off, and perhaps because I had something else of value to offer.

I wanted to stay at that university in the Southwest, but my position was terminated after the one year I taught there as a visiting assistant professor. People told me that being let go from the university was not a personal rejection, but it certainly felt like one. Each university that turns you away is like a home that turns you back from its door. I thought that eventually I would have to assess whether academic life was worth it. I did not want to keep trying for something that was not going to reward me.

Invisibility

We are sitting in a circle. It is an evening class in a professional graduate program at a California university. The title of the course is Seminar on Personnel Relations Research. Unofficially, it is a course on interviewing. The students do outside interviews and observe and analyze them throughout the quarter. I give assignments that change the interview situations, topics, and methods but that focus each time on the interviewer's assessment of self-other dynamics in the interview setting. We discuss each student's interviews in class each week. In the discussions, I ask questions: "What did you do, how did you feel, how did your interview go? How did interview (a) compare with (b)? Why do you think they worked that way?"

I think of the students in my class as research subjects (to avoid thinking of myself as a teacher?). The students' experience is my data, much as when I conduct a study of my own. From their experiences doing interviews, we develop the body of knowledge that is the subject matter of our course. This is something I like to do: make the subject of a course not objective and external, not something outside our circle, but a product of probing each individual's experience.

In the last session of the course, the students feel good about what they have done during the quarter and they say they have taught themselves, without a teacher. I feel suddenly absent. I tell them, with anger in my voice, "If I had not been here . . ." I feel a need to be appreciated, just as they need to appreciate themselves.

My teaching thus is dogged by many of the same problems as my research and writing. If I teach invisibly, for instance, drawing my material from what the students say, reflecting them back to themselves, sooner or later I have to speak of myself. I have to say, "I am here," not because anyone asks me to, but because the situation becomes intolerable for me if I stay hidden.

Writing Personally

It is six years later. I have just finished teaching a course on women and organizations at another California university. The students in this class concluded, toward the end of the quarter, that they liked reflecting on their experiences and speaking of their emotions in class discussions and papers. However, many found writing of their feelings hard to do. They said they had learned to write impersonally in school and that the habit was difficult to unlearn. Two of the three doctoral students in the class felt committed to trying to speak more personally in their dissertations, but they worried about whether others would accept them. I was concerned about whether they would stick to it. Perhaps I saw members of the doctoral class with whom I graduated falling by the wayside, giving up individual principles in order to conform to common academic standards. Perhaps I wanted to be less lonely: I wished these students to do work like mine.

I was surprised by the appeal of a personal approach in this class. At the start of the quarter, I still felt scared by my experience of twelve years ago, when a student challenged the ethics of a self-reflective assignment. In part because of that experience, I did not require that the students in this class speak of their feelings in their papers, except on one occasion. I just kept suggesting they do so, and many did. In the end, many of the students wanted a self-reflective approach presented as a more upfront and formal part of each of their assignments. They felt they would have benefited all along from an emphasis on learning from their feelings, not just in the one paper I assigned that way (a paper on their responses to *The Mirror Dance*). In one of the final papers submitted for the course, I asked the students to list the five readings they valued most of the forty we covered during the quarter. The most personal readings were the most frequently mentioned. I had not expected that. I had, in fact, worried about how the students would receive the more personal writings, such as one by Nancy Mairs about a mental hospital experience, and one about her relationship with her daughter.[5]

Throughout the quarter, I talked in this class, more than I am used to, about my own experiences. I wanted to set an example, but it is also what came out. When I started to make more abstract comments, I tended to get lost and to trip on my own words. Although it was difficult because of the exposure, after a while I thought I was saying something worthwhile in class discussions only if I felt a rush of warmth and feeling and as if I were blushing while speaking. To feel this way meant that I was speaking from my emotions, as opposed to cutting them off. This way of speaking soon became something I felt I had to do, as if the class expected it of me. I certainly had led them to expect it of themselves.

Some of the students in the class were exceptionally good writers. They wrote flowing, skillfully crafted papers that were verbally facile. Two who did so wrote final papers in which they chose to drop their glibness. They were aware that they could produce papers that were verbal works of art where everything was neat and symmetrical, but instead they chose to write more roughly, in a way that at first seemed wrong to them and that felt harder: "This essay is late, in part, because of a lesson I learned: beware of whatever looks easy, too neat, because that's not how the truth is. My first crack at this assignment came out too manicured; it didn't ring true. I knew if I took the time to dig a little deeper, I would get to the complexity."[6]

Other students, too, felt their writing was wrong when it was personal. One said a friend of his told him he could not hand in a paper like the one he had written, because it was organized according to his feelings, not his thinking. However, I liked it when the students wrote personally and roughly and in a way they thought was wrong. The roughness issue was not new to me, but it seemed new. I, too, try to smooth things over in writing: to make the surface flow, to protect myself by perfecting my prose. These students reminded me that it really is not better to do that. It is better to keep the rough edges in, or more of them than I usually allow.

In the students' papers, I came across statements about the value of each individual's particular experience:

> I learned that what I have to say is valuable in and of itself and does not need to be generalized into obscurity.[7]

There were statements that weighed the fit between personal writing and a more traditional academic approach:

> The readings that spoke honestly and from experience had great power to move me and make me think. I used to think this self-reflective style of writing was useful for expressing an idiosyncratic point of view and that a more academic style was necessary for conveying any more general understanding of a social phenomenon. I no longer see these as a dichotomy.[8]

The students spoke of a difference between what they had been told was challenging and good in academic work, and what they now found:

In writing my own self-reflection, I experienced how this style of writing or discussing was not "easier," was not an avoidance of the more "rigorous" exploration of a "theory." . . . I recognized how much I had been co-opted into a male style of presenting theory and data (the supposedly hard work), a style that removes the author's voice, even when it might be most informative and revealing. Once again, women's work was undervalued. Women are better at self-reflection, or at least more willing to give it a try and more interested in hearing others' self-reflections as well. This was being called easier, when in fact it is much harder.[9]

The papers spoke about authority:

I had always assumed it was necessary to include the words of an authority figure in my papers in order to gain legitimacy. In addition, I actively ignored or tried to overcome my feelings toward my research. I viewed my feelings of fear and anxiety toward my research as stumbling blocks to be kicked aside.[10]

Resolutions of the various elements of writing personally for an academic purpose were not simple in any of the papers, but the students clearly spoke of an awareness of something different:

It has been important for me to read work written by women who take their own experiences as the truth.[11]

It seems that my feelings get subordinated and suppressed by some mechanism that I have developed that is so automatic that I am not even aware of it.[12]

The words I have written seem very awkward to me. I have never written a research paper like this one. Everything seems to have bypassed my head and come out right from my heart. It was a difficult thing for me to do—not so much to tap into my emotions instead of my logic but rather to write a research paper in a way that violated everything I have learned about writing academic papers.[13]

I think I've experimented in this class with speaking in the "mother tongue," but now I'm not sure what to do with it.[14]

In commenting on how it felt to write of their feelings, these students posed a challenge to me. I tried to tell them about it in our last class session, but I do not know that they heard. I, too, have been schooled in the regular way, so that even when I speak of making things personal, and of identifying emotions in relation to readings or research, I am reluctant to do so. Only when I have already made a mistake and gone in the direction of too much generalization—such as when I bury myself in other people's voices (e.g., *The Mirror Dance*) or write prose from which I feel alienated (e.g., "Fiction and Social Science")[15]—do I bounce back and state things more personally. The students in this class seemed to grasp a more personal approach to writing social science and to believe that it was valuable. The message for me was that I had better practice such an approach myself and not keep running from it. I should not be afraid to assert what, for me, was a necessary way to speak.

I was reminded that when I wrote *The Mirror Dance*, I did not include a personal statement in the book but submitted one for separate publication later. That was also the case with *Hip Capitalism:* the study and a personal account describing how it was done were published separately. In my "Fiction and Social Science" book manuscript, I alternated authoritative statements with personal ones in the same text. When critics thought the personal statements were unnecessary, I shelved the document. However, I felt I could never again write social science without writing it personally. This was less a matter of principle than of development: writing academic prose in a first-person style was something I had grown to want to do. Because it seems right to me, I think I underestimate the discomfort such an approach can arouse both in others and in myself. One student in the class on women and organizations wrote of this discomfort when speaking of my article "Beyond 'Subjectivity,' " which provided personal background for *The Mirror Dance:* "A part of me was afraid for her. It was the same fear I had for myself. At times, I literally cringed while reading, thinking, 'Don't say that! There will be people who read this who will not respect you for it, and you've made yourself vulnerable to them.' "[16]

To write in the first person, as I have been speaking of it, is to include one's feelings and to risk exposure. It requires constantly asking the question, How can I best describe myself? And it requires an attitude of kindness toward the answers. It is helpful, in using a personal form, to understand the way specific experiences speak to one another: not through a style of generalization, but through resonance and articulation. One person's experience elicits an emotional response in another, for instance, because the experience is presented emotionally and in specific detail (resonance). One person's experience plays out aspects of experiences that others may share, but others may feel these experiences in a less defined form (articulation). To write of experience using a personal approach is not to choose a superior or inferior academic style, one that should be forbidden, or one that precludes speaking of others. Nor is writing in the first person the only way to produce an honest and self-involved account.[17] A personal approach is simply one way to model experience with responsibility and awareness. It is an approach we often leave underdeveloped because of our fears of exposure and because of an assumption, largely unsupported, that individual personal expression is merely self-indulgent and not more broadly relevant.

Often, I wish I could write more candidly than I do, with less resort to camouflage and generalization. That recent class seemed to say to me, "Try." It was the first time I had felt teaching so completely circle around. I had taught the students to experiment with something I valued, and then they expected no less of me. A student in a military uniform once tried to tell me that students wished to learn from teachers, rather than from textbooks. A woman in a seminar on qualitative methods made me think about the sensitive nature of asking for personal disclosure in an academic setting. This recent class pushed me to acknowledge more fully what I was about, if only because my concerns affected them and raised questions that were sometimes difficult to answer.

I still do not feel much like a teacher. Personal relationships are stressful for me, and teaching seems to require them, even at advanced levels. My old friend in the overcoat is still around, as persuaded as ever that teaching can be a matter of well-protected pedagogy. I feel exposed compared to him. Would he have listened to my objecting students if they addressed him? I think so. He would have wanted to know what was going on. I do not dress like the old man, and I act differently, but there continue to be two of us entering every classroom. The professor sits back from the table and looks around. I, very nervously, plunge in and speak.

In one sense, the duality I experience in classes reveals an observing ego and an active, risk-taking one. In another sense, it suggests two different ways of knowing the self. One way uses categories (the image of the professor). The other way speaks of a self that escapes the image.[18] It is this latter self that personal writing seeks to express, and it is precisely for this reason that personal writing is hard to accept in social science. Social science is valued for its way of understanding unique and particular phenomena in terms of more general processes. In identifying those processes, social scientific authors are expected to speak in a version of a standard voice and to understate their individual differences. For a writer to ask to be listened to for other reasons—because of a particular vision based on uniquely felt experience—is to ask for a different kind of attention than social science usually gets. Nonetheless, it is exactly this kind of attention that seems important to me.

12

Snapshots of Research

To us, Pauline, who is sparing with words, said after clear-
ing her throat, "Offer your experience as your truth." There
was a short silence. When we started talking again, we
didn't talk objectively, and we didn't fight. We went back to
feeling our way into ideas, using the whole intellect not half
of it, talking with one another, which involves listening. We
tried to offer our experience to one another. Not claiming
something: offering something. —Ursula K. LeGuin, Dancing at the Edge of the World

Our Stories

IT WAS 1972 and I was doing the research for my dissertation,
which would become the book *Hip Capitalism.* I was caught up in
my project and self-conscious about myself as a researcher. Often
my worry took the form of a concern with how I looked, as if I
would be accepted or rejected on that basis. In appearance, I did
not stand out or fit in. I felt undefined. I did not dress in hip or
business clothes like the radio station people I was studying. I
wore the same two pair of pants and a sweater most of the time.
My hair did not do anything special. I was not obviously a
lesbian, which might have distinguished me. I thought perhaps I
had a face that people felt comfortable talking to.

When I interviewed people, I felt invisible and awkward
and that I did not matter. Nearly everyone I asked talked with
me. I listened and took notes and saw myself as a recorder. I
looked at old pictures of the radio station staff, gathered docu-

ments, copied everything I could, and amassed a collection of pop books on rock music. I wanted intimacy, although I did not know it, and was uncomfortable when people gave it to me, when they revealed details about their personal lives. I was moved by the wrong people, such as those who were in managerial positions: the conservative corporate accountant; the marginal small-businessman who was afraid to come into work; the Big Daddy of the station. Whatever my emotions, I tried not to show how I felt other than to reveal a positive attitude toward my work and toward the people with whom I met.

The period of my research was full of tension and anxiety: Would the program director talk to me? If one person did not like my questions, would word get out and would anyone else talk to me? Would people answer my calls asking for interviews? If people answered, would they say yes or no? If they said yes, would they then cancel? Would they call to let me know if they wished to cancel, or would I drive all the way to their houses and find them not there, or not willing? I hated to pick up the phone. That was the hardest part. I kept putting the calls off. I had to badger myself into doing them: "No call, no interview, no research, no book." This sequence was perhaps the only element of my study that was linear. It was also hard to leave the interviews, to break away from the warmth of people talking about experiences that were important to them.

I used to think about the dynamics of my interviews as they occurred, marveling that people would take me into their confidence and wondering why they did.[1] In writing up my study, although I had other sources, my interview notes seemed to me the most alive aspect of my research. As I read, reread, and memorized the notes during the process of writing and revising from them, I repeatedly remembered each person I had met. Now, seventeen years later, I can still see, in my mind, many of the interview settings: a house in the Marin woods with a sunken bathtub; a downtown San Francisco bar; a funny curtained living space in an outer city district that was reached by climbing up a ladder; the back room at the station.

I cannot see faces well when I recall the interviews, but I can feel how the sessions went, or, more accurately, I can feel my own awkwardness. I see other people's bodies, types of hair, moustaches, remember key phrases, stories, meanings, things that mattered to each person, but I feel my own fears. What would the Big Daddy think of me? Would the night disc jockey know how much I felt for her? I studied the people of the station as one might insects, getting a sense of their characters, trying to pin them down in my mind to make them predictable for me. At the same time, I felt deeply for them, fell in love with some of them, wrote poetry to them that only I read.

I was researching a book that would tell the radio station's story. This book was supposed to matter to the people of the station, but probably it mattered more to me. The eighty-eight people I interviewed, and others they spoke of, still affect me a great deal. I have not gotten them out of my emotional interior. They come to mind at odd moments. I will suddenly remember a phrase from an interview and recall the person who spoke it and see the page of my notes on which it occurred. I will watch "Lassie" on television and think of one of the station's salesmen who once tried out for the boy lead in "Lassie." I will visit Marin county, or read about Los Angeles, and think about people from the station who moved there.

I am saying that research is a process that affects the researcher most of all. There have probably been few, if any, effects of my radio station research on the people I studied. The two most likely consequences are a momentary sense of having been important (important enough to be interviewed and to be the subject of a book) and a longer-term sense of having been sold out, having been used for the researcher's ends and not one's own. Once when I was teaching a research course using *Hip Capitalism* as a case study, I spoke of feeling let down because the people of the station did not respond much to my completed study, either in its distributed dissertation form or when it was later published as a book. A student replied: "They each told you their story when you interviewed them. They told you how they saw things, and they thought you were seeing it their same way. Then you made them pinheads in your book. It wasn't any one person's way. It wasn't everyone's way. It was your way. Of course they would feel betrayed and they would hide it from you. That's why you did not get the responses from them like you expected."[2] I had thought I knew that. I had read Joan Didion's comment that "writers are always selling somebody out."[3] However, I had wanted to do better.

At the time I wrote *Hip Capitalism,* and even later when I wrote *The Mirror Dance,* I had an essentially populist (power to the people) attitude toward doing social scientific research and writing. I believed that I was taking the people I studied into myself and that I would later give them back to themselves. I also thought they would be grateful.

I think now that I underestimated my own influence on my work and the likelihood that the people I wrote about would not appreciate my descriptions of them as much as I did. Although I wished each of my studies to be for the people whose lives my research drew upon, each study was of necessity, first of all, for me, and its final imagery reflected that. I think many of us who do social research want to give something back to people. One way to do this is to work at grasping our subjects' worlds and rendering a version of their reality that we believe to be faithful. Because this effort is often conscientious, it is difficult to concede that the people we study largely escape our descriptions of them, that the stories we write are always ours more than theirs, and that this is not a bad thing. It is not bad because a person who escapes a description, or who says that a social scientific account is more about the social scientist than about the person studied, is essentially saying to the social scientist, "I stand free. I have seen what you've written and it is not me. I am more, or better, than that."[4]

If social scientific accounts do not ever adequately represent the people they are about, and if these accounts do not work well as repayments to subjects for having shared their lives, how do they work? My sense is that they work because people who let social scientists study them are giving an altruistic gift. They are contributing to the development of knowledge, not knowing where that development will lead or exactly how it will impact them. They believe that some attempt to bring things out into the open, and to clarify them in more than one mind, is better than no attempt. They may have other beliefs and hopes as well (self-aggrandisement, for instance), but my point is that social science, at its best, is a gift of freedom. Any study of mine will be only loosely linked to the people it is about. It will be more tightly connected to me.

Sensitivity to Exposure

During the year I spent in the midwestern lesbian community on which *The Mirror Dance* was based, I worried about whether I would be kicked out or let in. I pretended I was apart from what I was not apart from: the community, the Midwest, a need to belong. The themes of my life became, in the end, the themes of my book, but it took me three years to identify them. Elsewhere, I have written about the difficulty of not being able to deal with the data of my study (see chapter 13). Here I want to speak of responses to the published work and to its very similar prepublication form.

Generally speaking, I received more responses from people who were not part of the studied community than from members of it. Some of the outsiders told me that *The Mirror Dance* reminded them of their involvements with other lesbian communities, or with other groups in which they saw similar dynamics. Others said the study probably reflected only its author and the particular community written about.

What of the responses from community members? After completing the manuscript of *The Mirror Dance,* I sent two copies to a core member of the community and requested that she circulate the copies to others in the community for comments. Within a few days, she phoned me long distance. She was very angry. I had violated everyone's privacy by writing this manuscript, she said, and she wanted to return it to me rather than circulating the copies further and adding to the damage.

I tried to defend myself by saying I had abided by the requests each person made at the time of her interview concerning use of her interview material. My caller said that another woman in the community was planning to sue me. Even though I had changed all the names as I had promised, everyone in the community knew who everyone was in the manuscript. They were, therefore, all exposed to each other. A few of the women had drawn up a list matching my fictitious names with their real ones, and they were sure it was correct.

No one, my caller said, when asked for her preferences regarding use of her interview material, could possibly have guessed that I would do "this" with it. I had not provided any new insight. I had simply taken what the women of the community said and quoted it. It would have been better if I had turned them into statistics. Then at least the community would have learned something from my study.

None of this made sense to me. On the phone, I tried to argue in my defense, but it seemed pointless, so I tried to quiet things down. I felt I did not want trouble. I had wanted the women of the community to be pleased by what I had done, but if anyone was to be pleased—perhaps the marginal people (those not in the core group)—now they would not have the chance to pre-read the manuscript and let me know. I was not going to send the manuscript back to the community after it was returned to me. I did not want a group of women from the Midwest calling up my publisher and declaring that my proposed book was unethical because it was a violation of their privacy. I did not want to become an issue in the lesbian press.

After my phone conversation with the community member who called, I wrote a letter trying to reassure her that my manuscript would not do harm. It probably would help other people understand themselves, and it would make them grateful to the members of this community for serving as the site of the research for the study. In preparing my letter, I decided that internal community exposure was not the key issue behind my caller's discontent, although she had said it was. The central problem, I felt, was that members of the community feared exposure in the outside world. They feared what would happen to them if the internal workings of their group were revealed to others. Others might learn from their experience when it was published in a book, but it would be their dirty laundry that would be trotted out to enable this. As a consequence of being revealed, members of this community might be ridiculed, disregarded, or judged in negative terms by others who did not understand them.

I think there is a parallel between fearing exposure of a group one is part of and fearing exposure of the self. Similar feelings are raised. The two are also connected. When a group one belongs to is exposed, the individual member feels exposed because the group reflects her actions. She may also feel exposed because an awareness that other people are watching makes the individual look more intently at herself in order to see what they see. It is at this point that inner fears and judgments arise, which may then get projected outward, making the situation doubly uncomfortable.

I went ahead and submitted my manuscript for publication. Whatever the community members felt, no one contacted my publisher or wrote angry letters to the lesbian press. No one from the community wrote to, or called, me. The moment of saying "I do not want to be seen" occurred in a single phone call late one night. That call was properly part of my research and, more quietly, part of my education. It did not signal to me that my manuscript ought to be stopped or changed in any way. However, it did suggest a sentiment that accompanies it.

That sentiment has to do with sensitivity to exposure. In writing *The Mirror Dance,* in the end of a chapter titled "The Outside World," I tried to describe what underlay the paranoia— the fear of being found out as a lesbian—that was felt by many women in the community. I strung paraphrases from my interviews together to get as close to a statement as I could:

> It was none of their business, and she just didn't trust them. She didn't trust people—the society—not to use that information against her. Why? What did she expect would happen? She was not afraid of losing her job. The fear was of being exposed, of leaving open a part of her that was none of their business, a part that was really personal, really private.
>
> She did not, said Nikki, think that fear of telling people you were gay ought to be overcome, because it was not the topic of homosexuality per se that mattered. It was the fear of having something too personal exposed.[5]

I do not think feelings about exposure should stop or change research. However, I do think they should be recognized. Research exposes people, whether one likes the fact or not. The exposures of research reveal tenderness. They raise questions of how we might create a world in which exposure does not result in hurt.

Toward Visibility

I had finished my *Mirror Dance* study and was ready for new work. I started interviewing residential real estate agents. I began to become familiar with local real estate prices and to learn about the selling of homes. I was interested in the interpersonal processes involved in residential sales—the intimate and emotional aspect of the work that women agents were especially good at. I was learning about these processes by listening to real estate agents talk about themselves. My sample was becoming clearer, as was the eventual form of my book.[6] I thought I would, in the end, write a text that bore a resemblance to that of *Hip Capitalism* and *The Mirror Dance*. It would be multivoiced and would examine the personal side of what is often viewed as only a money-driven business. It would make a statement about the meaning of home, privacy, and professionalism. My interviews were going well. However, I was haunted by a sense that I did not really want to be conducting my research.

I think I knew vaguely that I was setting up my study so that I would again become invisible. My invisibility would be encouraged by a research process that attended primarily to what other people said, and by a writing process aimed at speaking through others' voices. My experiences and feelings would be involved, but they would not be the focus, or the point. I thought this was the way to proceed if I was to do well. I would be drawing on my previous research experiences and doing something I had already learned I could do.

However, my problem with invisibility was worse in the case of the real estate study than it had been in my prior studies. The real estate agents I spoke with were highly oriented toward success, and I was interviewing them at a time when I felt very much a failure. I was soon to be out of a job and, although I had faced unemployment before, this time I did not feel I had an academic future. In my eyes, I was not a legitimate person. The world of real estate agents, like the world of aspiring academics, seemed to me, right then, to be one in which money mattered, in which "making it" mattered, and in which being busy in response to external demands was a sign of status. My busyness was of an internal sort. I was not a professional in the same way as the women I was researching: I did not place a high value on being businesslike or seeming competent. More critically, I felt inferior to these women. They were making new lives through their work. I was doing a study in an attempt to stay in place. I had not chosen the subject or style of my study because it would fulfill me, but because I felt I had to do something like it to keep going as an academic.

When I finally decided to stop the study and to give up interviewing real estate agents, I felt like a failure. I tended to blame my failure on a lack of institutional support. Had I a job or a grant, I thought, I would have kept on. Indeed, funding might have helped me, but I cannot say surely that it would have helped. I decided that if I had to pay for my own work, I would write a novel pertaining to my life. I wrote one and titled it "Jenny's World," Jenny being me.[7] Although the first person would have made more sense, I drafted the novel in the third person because I wanted the novel to be publishable. I decided to be cautious in discussing internal emotional pain in my novel because I thought people did not want to read about pain. Nonetheless, I liked the work. It was a way to be in touch with my own experience, and especially with feelings involved in prior lesbian relationships. It was also a way to bring others into my inner emotional world, both by imagining the people I was writing about and by thinking about potential readers. Bringing others into my inner life was important to me because I often experienced my emotional world as isolated and unacceptable. In writing my novel, I tried to emphasize ways in which my world was hospitable, a positive place, which is not to say that it was without difficulty.

In writing the novel, I did not have to talk to people to gain information, nor did I have to adapt to others' lives, making myself invisible in the process. Instead, I had to focus on my own experiences and to decide what I wanted to say about them. Yet the novel was not only about me, I thought; it was also about experiences of certain types that I felt existed in reality and that I thought people should know about and accept, and so know about and perhaps accept me. Clearly, I felt more directly present in the experiences I described in my novel than I expected I would be in my study of real estate agents. I also felt greatly relieved because the novel was work I was able to do.

I think my social science had become too remote at the time of my real estate research, and I had become too affected by a sense that I was not acceptable. Yet I did not know, nor could I imagine, how I might do social science any differently. I think I turned to the form of a novel in order to deal, in a more compelling fashion, with my own experience, even though I could not escape, with that form, all of the problems I confronted in my social science. It would take me several more years and another unfinished study[8] before I came to the recognition that I needed to be visible in my work. I needed to be able to speak of how I was present in it, and I needed not to understate my emotions.

On Interviewing

In all of my studies I have interviewed people. I have felt that interviewing is the only way to get the material I am after. For me, observation falls short as a research strategy because I feel that when I look at people, all I see is their external behavior. I do not know how they think or feel inside, or what their external behavior means to them. Library research, or reading, is not adequate because it seems indirect to me. I do read as I go along to supplement what I am finding out from people, but reading has never seemed like real research to me. I want to go out and get fresh new material, to feel that I am catching something live. When people speak with me in interviews, I always feel I am getting rich material. As my interviews pile up, or as one interview continues, I gain more and more. It is a very positive feeling of connection.

When I sit opposite, or beside, a person I am speaking with in an interview, I concentrate on what she is telling me and I take that person into myself. Later, my feeling for the person will guide me as I attempt to think about my material and to write up my study. I tend to think that interviewing is one of the best things I can do because it is a way of giving to other people. In addition, because I am a loner by habit, interviewing provides me with an opportunity for social contact that I very much need. It is a way of having intimate relationships and of structuring those relationships so that they will be safe for me. They will be compelling, about something central and strongly felt (by both the interviewee and myself), and they will be guided so that they do not get out of bounds.

Why, then, if interviews are so good for me, if they provide so much valuable and nourishing material, do I hate to interview people? Why do I find interviewing a hard experience, one that has gotten harder over the years, so much so that I now wish not to do it anymore? The answer to this question, for me, has to do with alienated and unalienated research. It has its source in a difference between the ways I have taught myself to do research, the rules I have thought I should follow, and personal needs that are not comprehended in those rules. In conducting interviews— in the radio station case, in the lesbian community, with the real estate agents, and in more recent work on academic women—I have sought to conform to common ideas about what an interview ought to look like.

In each study I have felt that the person I am interviewing should do the talking because an interview is supposed to be other-oriented, or other-serving. As I sit across from another person in an interview, in her kitchen, at a table, microphone on, taking notes, I say to myself, "The interview has got to pay off for that person in the quality of attention she receives from me. Later, you will have your feelings. Later, you will be able to forget how painful this interview experience was. In the quiet of your study, when you sit down to write, you can say whatever you want to. You will have the last word. For now, sit here and be quiet. The other person is the important one. Listen carefully to what she says. Make sure you have enough good material. When the interview is over, you can take it and disappear. You can make off with it like a thief in the night.

"For now, if you absolutely have to, you can say something to this woman about what you feel or think, but keep it short. Say your feelings or your thoughts at the end of the interview, or insert them somewhere when she pauses for breath. But remember that she does not really want to listen to you. That is the point of the interview. The hidden agreement is that you listen and she speaks. If she is giving you all this good material, the least you can do is to flatter her, make her feel important, central, fascinating. She may not like what you say in your study ultimately, but then that will be your business. The interview process just has to be like this. It cannot go both ways."

I have said some version of these words to myself many times. I have sat through many interviews intensely interested in what I was being told. All the while, I have felt terrible about myself because it was not all right for me to speak, or to be part of the interview process in a reciprocal way. The process, as I set it up, was one in which I allocated roles: listener/speaker, central person/admirer, woman with the goods/the thief. Interpersonal relationships, however, are not clearly divided in these ways, and what my splitting of them did was to force underground my part of each interview relationship. This produced a situation in which I then felt slighted and hurt because I was denied. Pretty soon, I was running from each interview, thinking only of the relief that would come later. I was present while talking to the other person in the interview setting, but in my mind I was frequently gone.

My splitting of my interview process into roles seemed necessary at the time of each study, not so much because I thought interviews had to be, by definition, other-oriented (by definition, they are inter-views), but because I thought interviews simply would not work for me if I acted differently. If I did not carefully split the interviews into roles, I would find myself entering into very messy relationships, unclear relationships, in which I would have needs. These needs would probably not get met, and then I would be hurt. Better to leave the relationship formal and short, I felt. A messy relationship could take a long time; it could be like a tortuous love affair. What if I told someone I interviewed that I loved them? What if I never wanted to leave them? What if I wanted to be their friend? (That they should want to be my friend, and try to be, was understandable to me, but the reverse seemed unconscionable.) I was a person whose needs, I felt, could not be met, involved in a process that, if it became reciprocal, would be morally wrong. Thus, I adopted a very alienated and painful strategy of interviewing. I did not change my strategy when it hurt me.

I simply repeated it. I split the roles and escaped, split the roles and escaped again, through several studies.

Now, not all people who do interviews experience their interview processes as painful, even when they split the roles as I do. Some people find interviewing enjoyable.[9] My point is not that interviews ought to be conducted in any particular way by everyone. Rather, I wish to say something about patterns of research. I think it is very possible, and common, to engage in alienated habits of research and to feel that one has no alternative other than self-denial. A tendency toward self-denial occurs not only in interviewing, but in all types of research: in observational, archival, and quantitative research. Tendencies toward self-denial are a consequence of disciplinary trainings that encourage separation of researcher and researched, and they are a consequence of desires for control, superiority, rational understanding, or for managing one's research relationships with a good conscience.

The alternative to self-denial in research often seems to be a path of "mucking it all up" as a result of one's personal involvement. A researcher worries that seeking to have her personal needs met will lead to bad outcomes in terms of her research. For example, her needs will bias her research disproportionately in favor of herself (she will display ethnocentrism or subjectivity), or her needs will exploit others. The basic concern is with the possibility of creating a distorting picture of the world as a result of meeting personal needs through research. Yet questions can be raised: Isn't the entire scholarly process a meeting of needs? Why censure the most obvious signs of that? Why are some needs legitimate and others not? Who determines this? Which inner voices ought one to attend to in thinking about one's needs in research? How can one best cross the line between self and other? How does one feel compelled to cross the line? When is pain a signal to pay attention to? How should one respond to one's own emotions?

When I think about my experiences of interviewing, I always remember the pain I have felt. Usually, I am reluctant to speak of that pain because interviewing is normally spoken of, by those who do it, as a rewarding and pleasurable experience. I feel that if interviewing has been painful for me, that fact is either idiosyncratic to my personality or it is a personal failing—a failure to conduct my interviews well, or to be a happy enough person. I tend to think my pain is largely irrelevant as a methodological consideration if the results of my interviews, the material I have gained, are rich and valuable.

My interviews have certainly been valuable for me. They have helped me to write three books. They have given me people for my inner life: each research process is like an internal home movie for me. When I read over my studies, I see replays of interviews. The books and the people in my inner movies are important to me. Nonetheless, somewhere inside me there is a small voice that refuses to change its tune. That voice says that self-denial, for me, is never a good thing. No wealth of material gained from an interview is worth the pain of a process that repeatedly proves to me that I can be a machine, that people will talk to me, and say good things, but only if I keep myself out of the interactive process, or if I keep myself out of it so far as they can see.

I have not yet solved the problem of how to do interviews. Perhaps, temperamentally, I have done as well as I can. I may be particularly sensitive to the opinions of others, particularly in need of reciprocal arrangements, and particularly reluctant to take the steps necessary to make such arrangements possible. But I may not be unusual in these ways. I know I feel that if I deny myself in interview settings (if I put myself on hold or keep myself out), and if I then base a study on my experiences in those settings of denial, I come to resent the people I have interviewed. Sitting talking with someone else about her life, when I deny myself, I feel, all the while, that she is denying me. The other woman is the one not hearing me, or unwilling to listen. Probably, I think, she is incapable of listening. She is concerned only with herself. "As is her right," I then add. Then later, when I listen to the tapes of her speaking, or when I read my notes, I feel bad and I hate her because I feel she has silenced me. I also feel all mixed up. I still want this other person to like me, in part because I like her, in part because I have become so involved with her. I hate that she does not know me. I hate the feeling of disapproval I have, or the feeling of lack of approval, on her part and mine, the feeling that some needed response is missing.

I am not sure, but I think it would take a superhuman act of love to keep all these feelings from entering any account I write. In the end, I write. I try to understand in the most compassionate way I know. However, I think my experience of denial makes it much harder for me to understand anything or anyone. It also makes it harder for me to write, or to articulate what I wish to say. I would prefer to operate within a system in which my needs for reciprocal understanding were met, and in which my research was not premised on an assumption about the desirability of cordoning off the self.

13

Beyond Subjectivity

IN EARLIER CHAPTERS, I have referred to an article titled "Beyond 'Subjectivity': The Use of the Self in Social Science." I have mentioned that the article meant more to me than the book whose research and writing process it comments upon because the article speaks in a first-person voice and feels closer to me than the book. I have mentioned, too, that when I began writing *Social Science and the Self*, "Beyond 'Subjectivity' " was very much on my mind. It represented a kind of personal writing that I wanted my new work to live up to. I was afraid I would not be able to be as candid in my new work as I had been in that earlier article.

"Beyond 'Subjectivity,' " presented here, provides a specific discussion of research and writing in the case of *The Mirror Dance*. It also extends themes raised in previous chapters of this book. It is concerned with acknowledging one's involvement in one's work and with achieving some level of honesty in writing about that involvement. The article argues for use of insights about the self to help one understand others, and it advocates the development of a full enough sense of self, so that understandings of others will not be stilted, artificial, and unreal. Even more important in terms of the present work, "Beyond 'Subjectivity' " takes as central the problem of asserting oneself, and one's own vision, or voice, in the face of other voices that often seem to overwhelm and discourage the social scientific author. The most important feeling I arrived at after going through the exercise described in this article, which was aimed at helping me deal with my data, was that "I had a right to say something that was mine." I think that often we feel we do not have that right in social science, or we feel it is a right we have to earn. I think such attitudes toward the individual authorial perspective, while appealingly modest, are not very helpful. They encourage us to deny that we will speak of things in terms that reflect how we see them. The more important question, is How will we mold these terms? What resources will we use to make our language fit our experiences? Will we draw fully on our internal and individual sense of things? Will we learn how to make good use of ourselves, or will we primarily apply commonly held views because we know these are acceptable?

In part because it draws on notes about personal feelings that were not originally intended for publication, "Beyond 'Subjectivity'" gives a sense of an inner authorial voice. It is an inside story about a book and an author, however brief and well rationalized. (See chapter 5 on the "success" storyline.) The tradition for writing up personal accounts in social science says that our studies are about others, and that our methodological statements should describe how we came to know what we did about them. Dutifully, "Beyond 'Subjectivity'" does this, but the article interests me now less as an explanation of how I came to understand a lesbian community than as a statement in its own right that presents aspects of the private world of an author.

To take the internal life and make it external is important, in my view. The challenge of making a true portrait of one's experience remains, but at least the self is acknowledged. All of our statements about others are, very significantly, also about ourselves. We tend to provide little in terms of direct personal discussion when we write our studies, and I should like to see that little become a great deal more. Only with many stories will we get a good picture, since we each can speak only of our experience, and often we do this timidly, afraid of the outside world's tendency to deny us. This general tendency is well fueled by the many specific prohibitions against self-expression within social science. These prohibitions are particularly strong in their effects on the self-expression of women and of anyone not speaking a standard truth. The following article is not intended as a model of originality or correct personal expression, but it is a piece I found helpful, and I hope it suggests that there is more to do. Not only do we need to start talking about ourselves at greater length, but we need to experiment with different styles of self-understanding, for these can be keys to expanding our alternatives, both for being present in our works and for depicting experiences of others.

BEYOND "SUBJECTIVITY"

THE USE OF THE SELF IN SOCIAL SCIENCE

Recently, in both the social sciences and in related humanistic disciplines, there has been a restimulation of interest in the relationship between observer and observed.[1] Our attention is called to the many ways in which our analyses of others result from highly interactional processes in which we are personally involved.[2] We bring biases and more than biases. We bring idiosyncratic patterns of recognition. We are not, in fact, ever capable of achieving the analytic "distance" we have long been schooled to seek. While recognition of the interactional and contextual nature of social research is not new,[3] how we interpret ourselves during this new period of self-examination may, in fact, add something fresh and significant to the development of sophistication in social science.

I present the following account of my own work with the hope of contributing to a general sharing of personal stories about what we, as social scientists, now do. My account is one of backward beginnings, wrong ways of doing things, and problems I would rather not have had. Yet precisely because of these things, I think, the story is worth telling. In the following sections, I tell about some of what went into the writing of *The Mirror Dance: Identity in a Women's Community,*[4] a study of a midwestern lesbian social group I conducted during 1977–78. The book focused on problems of likeness and difference, merger and separation, loss and change, and the struggles of individuals for social belonging and personal growth.

I began my research unwittingly. I spent nearly a year participating in the community as a member without the slightest thought of studying it. The community was, for me, as for others, a home away from home, a private social world, a source of intense personal involvements and sup- portive social activity—a source of parties, dinners, self- help groups, athletic teams, outings, extended-family type ties, a place for finding not only lovers, but also friends. I had moved to a midwestern town to take a job as a visiting assistant professor and had found the community by acci- dent and through need. My participation surprised me. "I did not become a lesbian," I wrote to myself in notes at the time, "to become one of a community." Yet the community won me over in the end, and three months before I was supposed to leave the job and town, I decided to study the community in which I was living, to ask questions of these bold midwestern women.

Data and the Problem of Interpretation

I had, for years, been interested in the subject of pri- vacy, and I felt that this private, almost secret sphere of social activity would be a good place to talk to people about it. I wanted to learn about how individuals dealt with how they were known, or not known, to others. I then began two months of intensive interviewing with seventy-eight women who were either members of the community or associated with it. Someone joked that I had solved "the sampling problem" by interviewing the entire community "and then some," which was, by and large, what I did.

My interviews lasted an hour and a half each, were usually conducted in my own home, and focused on personal histories of self-other relationships in the community. I asked each interviewee four basic questions: (1) How would you define privacy (what images come to mind)? (2) How would you define the local lesbian community? (3) Within that community, how have you been concerned about bound-aries between public and private, self and other (i.e., what has been your personal and social history)? (4) With respect to the outside world, how have you been concerned about protecting the fact of your lesbianism (who knows, who does not, and why)? Approximately 70 percent of the time in each interview session was spent on question 3, which con-cerned internal community relationships. Members of the community and others I approached showed me unusual cooperation. They typically came for interviews within a week of when I called. During the interviews, they spoke to me with great candor.

When I left the community, I took with me, along with my personal memories and accounts, four hundred pages of single-spaced typed interview notes, which were, I felt, "rich data" for a study I would soon write up. Then the unexpected happened. For a year I could do nothing with my notes. I picked them up; put them down; moved them around; took notes on the notes; copied them so that one set could sit in loose-leaf binders in my university office while the original set lay in binders on my kitchen table at home. (I had moved the notes to the kitchen table after realizing I kept avoiding them at my desk.) All the while, I kept trying to do simple things; to isolate themes; to find some-thing to say that could be supported by my data. I thought of punching computer cards. I finally culled out the seven interviews with lesbian mothers and attempted to write about their experience, thinking that in some magical way the subject of motherhood would save me. It did not. Then I gave up. I closed the notebooks. I decided to write a novel. Occasionally, I thought about how despite the fact that I was now twelve hundred miles and many months away from the community, I still did not have a necessary analytic "distance" from the subjects of my study. However, that thought did not help.

Finally, a full year later, done with two drafts of my

novel, and haunted still by those volumes of notes—the undefined "promise" of my data, the sense that I should not let the research go to waste—I decided, "I must write about what I can relate to. I must write a personal account." I began writing about what it had been like for me to live in that lesbian community. I wrote many pages, and then I shelved them. What I wrote was interesting, to me. Beyond recalling my experience, it enabled me to see that what I had thought of as a lack of analytic "distance" might more usefully be viewed as a lack of personal "separation" from my data, from all those "other women's voices" that rose up each time I took up my notes. But the account I had written was not social science in a conventional sense, and I wanted very much to be conventional.

However, writing the account did give me an insight. The most immediate problem, it seemed to me, was not that I did not have distance from my data, but that I did have, and probably always had, far too much distance. Before dealing with problems of "separation," I had to acknowledge that I was estranged.

I thought about estrangement.[5] I decided that to deal with my data in any "sociologically useful" fashion, I would have to get over my estrangement. I would have to feel that I could "touch" the experience of gathering my data, and in a way that I had not allowed myself before. I would have to begin by expanding my idea of my "data" to include not only my interview notes, but also my entire year of participation in the community. I would have to be willing not only to feel again what the experience of living in the community had been like for me, but also to feel it as fully and deeply as possible and to analyze my feelings. Why did certain things move me? What had unfolded over the year's time? Why had I felt estranged? What did I want? What did I receive? What was I afraid of? How could I bridge the gap between myself and my data?

Because I am not at ease simply "feeling" in an amorphous way, I went about "becoming in touch" with my data very methodically, in a highly disciplined and structured fashion. For the next four to five months, I devoted myself to an exercise which I called a "process of reengagement." The first stage of this exercise was a step-by-step analysis of my experience of involvement in the community, beginning with entry, progressing through entanglements in personal relationships, singling out key events and my emotional responses to them, reviewing the interview period, and ending with my feelings on leaving. The second stage was a step-by-step analysis of my experience in conducting the seventy-eight interviews which were the source of my notes. I later wrote a personal account of this process, called "'Separating Out': A Method for Dealing with Qualitative Data," from which the following is excerpted. This excerpt describes the second stage in my reengagement process and shows how an understanding of the self can help resolve the problem of interpreting one's data.

"Separating Out": A Method for Dealing with Qualitative Data

A Case-Analytic Technique

The strategy in the second part of my reengagement process required that I deal with each of my seventy-eight interview cases. First, I sought to identify and examine my responses to my interviewees as individuals. I reviewed feelings I had with respect to each interviewee, first, in anticipation of our interview session and, second, during the interview itself. Finally, I analyzed the data of my interview notes themselves. The analyses in each phase of this process were done by writing down my thoughts and feelings, taking up a separate sheet of paper for each interviewee at each step. When I was through, I had one set of notes reflecting my "preinterview self-assessment," another on my "interview self-assessment," and a third on responses to the interview notes.

Step 1: Preinterview Self-Assessment

During this preinterview self-assessment step, I recalled my acquaintance with each interviewee prior to our interview and reviewed how each interview appointment had been made. I remembered social occasions during which the interviewee and I had met and what the biases in introduction had been if the interviewee was known to me primarily through another person. Most important, I noted my personal expectations with respect to each interviewee immediately preceding the session: what I had anticipated with fear, and what with excitement, and what I felt I had wanted for myself in return. In doing this, I sought to identify those prejudices I brought to each interview. It seemed important to separate my personal disappointments and pleasures from my latter interpretations of my data. The following examples are indicative of the preinterview item self-assessment. They are excerpts from longer passages written about each preinterview experience.

> PREINTERVIEW 32: B. was one of my neighbors across the street who had been fairly open and friendly with me. I chose her to do one of the first interviews because she had been "public" as a lesbian and I felt she would be knowledgeable about the community and straightforward with me. Yet I was nonetheless concerned that she might not speak personally enough with me.

> PREINTERVIEW 44: I knew D. mainly through K. and was prejudiced against her, or, more accurately, I felt fear regarding her—that she was judgmental and did not like me because of my relationship with K. and its troubles—that she was primarily K.'s friend.

> PREINTERVIEW 67: I knew of V. that she was a straight woman in one of the core support groups in the community. Was afraid she would be distant and would withhold. Also, K. had told me V. played "poor me," so I worried I might get impatient with her.

Step 2: Interview Self-Assessment

A similar approach informed the next interview self-assessment step. Here, again, I wanted to identify my prejudices and any "hidden" personal agenda I might have had. Yet in this step, even more so than in the preinterview assessment, I was intent on recapturing the force of my emotions at the time, since they seemed to me to surround my waiting notes. For example:

INTERVIEW 32: This surprised my expectations, because B. was, it seemed, candid with me and more personal than I'd expected. I did not feel forced to adopt her views or anything of the sort. I really felt for her as a person at the end of the interview, as I had not before.

INTERVIEW 44: Interview was very tense for me. I felt D. being defensive. Felt pressure on her part that I join her—see it all her way. Felt she didn't want to be interviewed, felt I was pushing this upon her. I was angry with her because of all these things. When I really wanted to be friends, to win her, to have her like me. In the end, I felt she ran away, wishing she'd not said what she did, angry with me. I wished to run after her, to make it all right—to confront. This is the interview I felt worst about of all of them, it seemed so much a denial and rejection of me. Though I felt its content—what she said—was rich.

INTERVIEW 67: Was partly tense because I suspected my own motives about wanting to get to K. through V.—get inside information that would help me settle my troubled feelings. Felt partly pressed by V. to feel as she did. Also that V. was partly confused, yet that she felt she had a collar on rationality. The conversation was almost technical, in that she kept much emotion out. Did not like this (angry?—a little, but repressed it) in the end.

INTERVIEW 72: M. was the only one to really break down and cry at the time of the interview and want to be held. This scared me—because I did not want to get involved, and did not want her to become dependent on me. I tried to "handle" it by not making a big deal of it, by holding her and then letting her up when I felt she'd be okay. She had brought a tape recorder to record the interview for herself (only one who did this). When she started to cry and asked to be held, I pushed the recorder off. My leftover feelings were fear—that she'd call on me for more holding and that I'd say no. I might do it with someone but somehow I feared her, or she was not the one. I also had feelings that I invited this, with everyone. Then when I got it, drew back from it. This left me uneasy. Feeling angry (?), lonely. What if it were me who wanted to cry and be held?

It became increasingly obvious to me, as I recalled and noted my responses in each case, that I had felt much discomfort and that that had caused me trouble. Yet these were exactly the kinds of things I needed to articulate, since they had been crucial in frustrating my dealings with my data. For instance, the more I noted my responses, the more I became aware of how very often I had been afraid, both prior to the interview sessions and during them. What I had previously identified as anger was really fear. This, I think, was because each interview situation was an intimacy situation for me and an occasion which I felt required proof of myself. I wanted, during the interview sessions, not only to know each of my interviewees, but also for them to know and care about me. I reacted as if it were a denial of myself when an interviewee did not seem to care:

INTERVIEW 62: A disappointment. Because M. seemed to me a lot a front—how she wanted to appear, a line, not a real person. I didn't feel the intimacy, the honesty that I wanted. Felt she suspected that I found her false (unconvincing) in this way and that she was angry at this and defensive. When she left, I was let down and angry. Felt she had dealt with me formally, almost as a functionary for herself, rather than as a person. Which I wanted.

The interview self-assessment was difficult. I kept wanting to describe the interviewee and how she appeared to me when she arrived and as she was involved in the session. However, this seemed largely ungrounded, unless I also noted my own reasons for the response—the emotional issue, or issues, each session raised for me. I had to discipline myself to note a reaction of my own for every action of each interviewee that I noted. I had to take time to figure out the logic of my own reactions, for what they would tell me about barriers to dealing with my data. I had not expected the interview assessment to become highly self-analytic, since I felt I had already been extremely self-reflective during the earlier stage of reengaging with the entire research experience (step 1 of my exercise). Nonetheless, new things were brought to my attention in recalling my specific feelings in each of the interview situations.

My most important recognition occurred after going through approximately one-third of the cases. I began to notice that I could distinguish my responses in terms of whether I had felt pressure to become like a particular interviewee or whether I had felt I could "be myself" during the interview. I then began to look to characteristics of the different interviewees in relation to myself in order to understand why I would feel or not feel pressure. I realized that I would become angry and feel bad in those cases where I felt I had to be like the interviewee. My sense gained from these cases tended to overpower my sense of the actually larger number of cases in which I did not feel this threat. For example:

INTERVIEW 75: Went well. I was impressed with R. as a person—her independence, the carefulness of her thinking, her clarity. I got a good picture of her—because her words were honest?—and so did not feel threatened. Maybe this is mechanism: when the interviewees are confused (due to being defensive or otherwise inauthentic, or confused about themselves), I get threatened and confused about who I am, because the relationship is confused. I don't know what I am relating to; while if the interviewees are more clear about themselves, I can be more clear about myself.

I concluded that I had felt threatened where it seemed to me an interviewee was inauthentic in her presentation of self in ways that set off my own doubts as to who I was. I decided that my feelings concerning this were so strong because of the fact that I shared an intimate identity stake with all the women I interviewed. I looked to them, even in the ostensibly other-oriented interview situation, to help me solve the problem of who I was. Although the interviews were highly controlled and guided by me, my controls did not protect me from threats on a deeper level. The interviews were actually occasions of inner panic, occasions during which I feared that others would not allow me to be myself—to act as the person I truly felt I was. This feeling of threat to my sense of self had not been fully clear to me before I analyzed the individual accounts. But finally it was, and I could see in my responses how much I had wanted personal confirmation and acceptance:

INTERVIEW 54: In her office. I was uncomfortable because of her power things—the phone, showing off stuff she'd written, her sensitivity. I felt she was trying to impress me with herself, that I was mostly a pawn to this, a person to be won over, not an independent person to be related to—one who had sensitivity, specialness, etc. And I wanted this other response from her. Perhaps because she was a peer at the university, and an unattached woman, recently out of a relationship. I think I had hopes we might be friends. With sexual possibilities maybe. But even if not, I wanted to be an equal, a real person to her. I left disappointed.

As I analyzed my responses in this way, I felt that my desires for confirmation, while perhaps extreme, might be more widespread in the community. The lesbian community might be functioning as an "identity community" for its members, one in which the most intimate sense of self was frequently on the line, a community in which the power to threaten by lack of confirmation was as strong as the power to confirm.

Reexamining my interview session responses made me aware of something else that was extremely important: the extent to which, even in those special-purpose sessions, I was engaged with the community and acting according to its rules, just as I had been outside the sessions. The interview situations were, in effect, small dramatic reenactments of the social dynamics of the larger community. They were microcosms providing specific examples of expected or acceptable community behaviors. In looking back on my responses, I was shocked, for example, to see how often my reactions to interviewees included an element of sexual expectation. In this way, I was clearly a member of the community:

INTERVIEW 2: B. was younger than I had expected and very beautiful, with long straight dark hair. She reminded me of a woman I had been involved with back in California in the winter. I think I felt I would like her to fall in love with me.

INTERVIEW 51: Knew E. casually, and occasionally, it seemed to me, she would be showing sexual interest in me. Some part of me, I think, wanted that more and also was repulsed and frightened by it.

The sexual expectation dilemma had been spoken of candidly by one of my interviewees:

> It's like good old sex being such an important part of people's life. And coming to a place with that expectation. Like I am here because other people in this room are here because they have the same sexual orientation I do. It puts a great pressure on you as to what am I up to and what are these people in this room doing? A lot of heterosexual traps I tried to escape, I found here. Because of all those sexual tensions, nobody gets to really know each other or to feeling comfortable with each other.

In this same vein, this interviewee also articulated a predicament referred to in the accounts of others:

> There isn't one woman in the community I haven't considered having a relationship with, just because you're in this community and because of all the pressure to need and want a relationship. Because you're in this community and because you have to relate some intimate details to get along, there is always the question of whether you want to be intimate. In a straight community, you have all these girlfriends who you tell things to. But in this community, you have to worry about whether it means you want to go to bed with them.

It was not easy for me or for my interviewees to acknowledge the pervasive and central sexual tensions of the community, since these were often subtle and personally sensitive. However, by examining my own responses, I was able to arrive at important insights. I concluded that my feelings of sexual expectation had less to do with actual possibilities for sexual relationship than with rules for defining the self in the community. For this was a community in which one's sense of personal identity was closely linked with one's feelings of sexual possibility and in which sexuality often appeared as a route to intimacy, as a means by which an individual might become truly known.

Step 3: Analyzing the Interview Notes

Once having completed both preinterview and interview self-assessments, and interpreting as many of my own responses as I could, I turned to the task of dealing with the content of my interview notes. Initially, I wanted to treat the accounts of my interviewees, as much as possible, as separate and different from my own. I wanted to see my interviewees as sharing my processes and reactions perhaps on occasion, but not as a rule. Yet as I began to review my notes, seeking concepts appropriate for categorizing and "making sense," I found that I was drawing on my understanding of myself with far greater facility than on anything else that came to hand. The task then reformulated itself as one in which I would seek to determine if and how my interviewees shared versions of the problems I had identified in myself.

I would look for words used by my interviewees that were reminiscent of my own, processes that were similar to mine, and assumptions about self in relation to others that were similar. Most centrally, as I read, I would imagine each individual as existing in a problem situation concerning differentiation of self in the community. I would view each individual as seeking, time and again, confirmation for who she was, all the while suspecting she might not belong.

Increasingly, as I examined the notes, I found what seemed to be parallels among the feelings expressed by the interviewees. For example, there was a frequent concern with possibilities for rejection by the community, whether rejection was or was not likely to occur:

> I have a yearning to be part of the community,
> but I feel, and I know by the grapevine, that I
> would be rejected.

There was a sense that the community had rules that excluded important aspects of the self, as these excerpts from different interviewees suggest:

> It's hard to capture because all that is implicit—a sense that the community does have these strong rules.

> There are some things you couldn't say.

A sense of the community as unreal or uncertain appeared often in the various accounts:

> The community, for me, became a monster.

> I see several different communities.

> Like the first two years I lived here, I was unaware there was one [a community].

> I think of them as a real tight closed group, that's closed until they know for sure that you're a lesbian, for one thing. And I don't think that you could just go meet them, go hang out with them. I think you have to join them.

In most of the accounts, there was a desire for the community to provide acceptance and self-confirmation:

> The community, to me, is a group of women who I could *know* that I could lean on for support.

> Here is a group of women who can understand me, touch me the way I want to be touched.

> When I had finally found these people, I felt I had finally found people who could accept my whole life.

Along with the need for confirmation were feelings of extreme disillusionment and disappointment when the community seemed to have failed a particular woman:

You would think it would be easier to assert your differences in a community of women. But it's not. It's real disillusioning.

I needed reassurance that I was doing all right. I needed some indication that I was appreciated. And they kept spewing forth this ideology of the community, the community, this axial of support when I felt totally abandoned.

During this data analysis step, I used my own insights and developed them further with reference to analyses of the interview notes. This enabled me finally to write a paper about the collective reality of participation in the community.[6] The reality I described in that paper was, of course, only part of the reality felt by community members, that portion I could be in touch with as a result of my experience. But by now, as a result of clarifying my experience, I was no longer as frightened of my involvement as I had been initially. I could use my own recognitions as a guide, a source not only of personal but of sociological insight.

This is not to suggest that my interpretations of the community were merely interpretations of myself "writ large" and imposed on the testimony of others. I also had to follow additional rules that granted to other members of the community feelings and responses that were different from mine. Throughout the process of analyzing my notes, it was important for me to maintain a sense that there was much in each interview account that fell beyond my own limited experience. My task was to try to uncover what I could with the tool of myself and my personal recognitions. I sought not simply to impose or to apply my newly developed recognitions, but to expand those recognitions by constantly challenging my existing understandings: challenging my perceptions of others with what I now felt I knew about myself and, at the same time, confronting my self-understanding with what my interviewees seemed to be telling me that was different.

I think that often in social research, this is what we really do. We see others as we know ourselves. If the understanding of self is limited and unyielding to change, the understanding of the other is as well. If the understanding of the self is harsh, uncaring and not generous to all the possibilities for being a person, the understanding of the other will show this. The great danger of doing injustice to the reality of the "other" does not come about through use of the self, but through lack of use of a full enough sense of self, which, concomitantly, produces a stifled, artificial, limited, and unreal knowledge of others.

Conclusion

The preceding account describes an exercise that helped me to reengage with my data at the same time as I was "separating out" a sense of myself. The exercise proved immediately useful in generating insights. However, my problem of estrangement was not so easily solved. In 1980 when I returned to California, I again had trouble dealing with my data. I wanted to work with it and, simultaneously, to leave it—to begin new research. At that point, it helped for me to think back on the exercise I had engaged in during the previous year. That exercise had given me some confidence and an initial understanding of the nature of my problem. I knew I would have to "assert myself," even if my assertion felt uncomfortable, and even if I would continually feel I was illegitimately imposing myself on my data.

Two and a quarter years had passed since the original research for my study was completed. I finally began writing *The Mirror Dance.* I wrote it quickly, relying upon what now seemed deeply imbedded intuitions. The book was published in 1983. Responses from both reviewers and readers suggested that its portrayal was valid, to a surprising and somewhat uncomfortable degree. Yet I knew that what I said in *The Mirror Dance* was dependent on a very personal and idiosyncratic process of data gathering and analysis. Because that process was so personal and because it worked essentially "backwards"—to understand my community, first I had to understand myself—I have presented a partial description here of an analytic exercise that helped me greatly.

The exercise I engaged in was, for me, a way out of a problem. It was a source of insight both about others and about myself. It gave me some confidence when I needed it; it gave me a feeling that "I had a right" to say something that was mine. I had studied a community that I felt I was part of and, at the same time, that I felt estranged from. I was, at one point, overwhelmed by the voices of all the women in that community. They were all telling me what to do, and they were each telling me something different. It took a long time—longer than I had expected—to find, in myself, a voice by which I could speak back to them.

I found that voice and, as *The Mirror Dance* attests, I hid it. *The Mirror Dance* is written in an unusual ethnographic style, in which the voices of the women of a community interweave with and comment upon one another, analyzing their collective situation. The subjective "I" of the author is hidden in the book, never mentioned, merged finally back in with the community from which it emerged. It is precisely for that reason that the preceding account seems important to me, for it speaks to the origins of the book's inner voice. More crucially, it speaks of a personal process. In social science, I think, we must acknowledge the personal far more than we do. We need to find new ways to explore it. We need to link our statements about those we study with statements about ourselves, for in reality neither stands alone.

V

Other Voices

14

Problems of Self
and Form, I

THIS CHAPTER AND THE NEXT are intended to function as an appendix and conclusion to parts I–IV of this book. They present the voices of eight women in the social sciences reflecting on the use of the self in their studies. The fields from which these women speak are sociology, psychology, history, education, anthropology, and communication studies. I present these eight women's experiences, gleaned from interviews, in order to show how others deal with the issues of self and form that I have raised in prior chapters. All of the women whose accounts appear here are concerned with the presence of an author in a work, and with the relationship between author and subject. Many of these scholars comment on the constraints of social science, and several discuss how they might write differently if they could follow their own internal sense of form to a greater degree than is conventional in their fields.

The scholars whose experiences I report on here are not a sample intended to be representative of a broader social group. They are simply eight women social scientists with feminist orientations. From the time I began *Social Science and the Self*, I wanted to interview other people to supplement my account of my own experiences. I picked women for my interviews because I felt they would be more articulate about the self in ways I was interested in than men would be. I chose women who had feminist orientations for a similar reason: usually these women had thought about self and work in a principled and extended way. Finally, these are people who were accessible to me. When I called and asked if they would be willing to speak with me, each quickly said yes. I spoke initially with ten women for this study and later decided to present the accounts of eight of them, because these were the accounts that spoke most directly to my concerns.

In starting the interviews, I told each person about the subject I was interested in: social science and the self. I asked each woman if she might talk about struggles in reconciling her own sense of experience with the forms of social science. I also asked each to discuss the use of the first-person singular in her work. These two topics (struggles with forms, and use of the first-person singular) seemed to me related because they both were about self in relation to work, and they both concerned the structure of expression in social science. The first topic, however, was more distant. By "the forms of social science" I meant especially the conventional forms: the way social science was normally written in each discipline. The second topic (the use of the first person) was closer to the individual. "When have you used the first person in your work, and how have you felt about it?" I asked. The interviews lasted one-and-a-half to two hours each. The women easily spoke at length, with little prompting from me. The subject of social science and the self seemed familiar to them, and it seemed important to talk about it, if only to work out their, and my, thinking a bit further.

I conducted these interviews after I had completed a well-defined draft of the first thirteen chapters of this book. I decided to conduct the interviews last not because I wanted to set them up to challenge or support my contentions, but because I was afraid that if I interviewed other people first, I would get so involved with what they thought, or felt, that I would have a hard time knowing my own feelings and thinking. I had had that difficulty before in *The Mirror Dance*, as I discussed in "Beyond 'Subjectivity,' " and I wanted to avoid it happening again. I also wanted to have my own questions and experience clearly in mind by the time I took to interviewing others, so that I could encourage them to speak of matters that concerned me. This required writing my book first. The practice of interviewing others last, after already forming one's own ideas, is not the standard one in social science. Interviews, and library searches, are usually done first, and then one develops one's own thinking based upon them. The practice of consulting others later has perhaps more affinity with the artist's quest to be original than with the academic's desire to come to conclusions based on evidence. To foster originality, one must often sequester, or protect, the self and its ideas and give them room to develop somewhat independently, without too much influence from others.

In the accounts that follow, drawn from eight of my interviews, I have condensed each woman's story, added a few words of my own to ease transitions, and focused on material that seemed to me relevant to my two central themes: use of the self in social science, and one's relationship to the conventional forms for writing social science in the various disciplines. In presenting these interview accounts, I have tried to grasp each person's experience and to be faithful to my interviewees' meanings and words. However, although faithful, the accounts are not verbatim, and they reflect, to a large extent, my own sense of what others have told me; thus no external quotation marks are used to frame a speaker's words.

These eight accounts discussing problems of self and form are presented individually in anecdotal fashion. Each woman takes a turn at telling her tale. None of the accounts is followed by explicit analytic commentary by me. My commentary is contained in the structure of each story. Subordinating the interview materials under topics did occur to me, and I tried it, but in the end it seemed preferable to place each speaker's remarks in the context of her individual train of thought. I invite the reader to make up her own themes as she goes along. This is material to be read through, listened to, and reread. The point is to take in the experience each woman describes, preferably uncritically. Then see what general impressions are left after reading the different experiences. What themes emerge? What is the feeling sense? What stays with you?

Interesting questions arise for me in reading these accounts. My main questions center on the following: What conflicts does a person feel when she thinks of writing differently? To what extent are these conflicts a product of her social scientific training? How does the individual think about herself? Why is she reluctant to use the first person? Who is she talking to in her mind when she thinks about what she can and cannot do? Why is it so hard to push away the prohibitions against visibility and self-expression, particularly for women? What do these accounts say about the process of creating works of social science, in addition to my particular concerns about the relationship between work and self? One gets a good feel for ways of answering these questions (and for the severity of the self-repression problem in social science) by listening to these eight feminist scholars. Their accounts provide many experiences that a reader may identify with, and identification can be an important route to learning.

Different interpretations may be made in response to these accounts. To risk one overview, though, it seems to me that the women who speak here have in common an early and continuing professional socialization in which social science is viewed as a very limited type of science, and in which the self of the social scientific author is viewed as a contaminant. Thus, they, and we, if I include myself, struggle with a sense that "This is what I learned. This is what they taught me (and 'they' are usually men)," and, "This is how I feel, and what I might want to do, if the prohibitions against being myself, expressing myself, or doing something my particular way, were less strong." The following accounts describe, in addition, what gets done in spite of the many prohibitions against self-expression in social science. In passive constructions, in the first person, using numbers or words, the women who speak here are authors who invest a great deal of themselves in their studies. Yet often they feel that what they know does not come through in their writing. The forms they have learned are not adequate to the content they wish to express.

All names and formal identifying characteristics of the women interviewed have been changed in these accounts. However, such changes do not protect the speakers against feelings of vulnerability. The women I interviewed did not expect to be quoted separately at length, for instance. I think they expected to appear in more cut-up accounts, as was the case in *The Mirror Dance*. My presenting their views in the form of individual stories may increase their sense of feeling revealed. My hope is that readers will interpret the accounts kindly and protectively, knowing that someone else may feel that, despite the protection of anonymity, her life is very exposed here.[1]

In closing with these eight "Other Voices," I continue to speak of my own concerns, which I have stated previously with a more direct "I." My goal now is to open my discussion back out to give a sense of struggles other than my own and a sense of where the issues lie in others' lives. My hope is that this way of closing can provide a base of multiple experiences to which a reader can relate. The struggle for self-expression is felt by many of us, and each deals with the issues in a somewhat different way. My particular way of thinking about the self in social science has much in common with the views of the women who speak in these pages. However, I differ from my interviewees in significant respects. For instance, I like to think of subjectivity not as the lens through which one sees the world, but as the world itself. This inner-world definition is, for me, the most important suggestion of my discussion of Georgia O'Keeffe. O'Keeffe's landscapes were not, for her, views of mountains seen in the distance. They were her attempts to articulate her own inner reality, which sometimes had pieces of mountains in it. I also think our studies are about us, and not primarily about others. Many of my interviewees would disagree. But my point is not that we should all agree. We can't. What we can do is to suggest the nature of our experiences.[2]

Other Voices

Myra: "I felt a strong emotional attachment to doing it this way."

I am now writing a book based on families in a high-technology industry setting. It is very different from anything I've ever done before. It is different for me. The research and writing are both different. I have been concerned with the use of voice, especially the personal voice, in this book. I did not have a sense of what kind of book I would write when I was doing the research. I set out to do a qualitative study of working-class family change in the area. Then I got interested in some stories of the women: their divorces, their feminism, events in their lives. I interviewed a woman who came out to me as a born-again Christian. I followed my nose and gradually evolved a case study. I wouldn't call it a method. Getting to know two of the women and their relatives and friends, and lives and histories, over several years: that was my method. It included formal interviews, and being with the women and their families, and being introduced to others.

This method was a departure for me. I had done library research before. It was also different from standard sociological procedure where there is a sample. These people adopted me. I felt seduced by them in a way. My research approach was different from standard anthropology too. I did not live there. My interviews and participation were intermittent, contingent on when I had leave time and when I was willing to leave my family. Thus, the research was different than I expected: no sample, two families (numbering thirty people), and more than two households. I heard stories that were like soap operas. There were tragedies. What in the world do you do with this kind of research? What kind of book do you write? It seemed to me the best way would be a novel. I felt immediately forlorn that I was not a novelist. I also wanted to be a novelist because of a wish for freedom. I wanted to be able to say what I wanted to, without violating people's rights, or leaving things out that they objected to.

I have written one complete draft. I tried at first to organize the book thematically. My initial fantasy was a novel. The thematic organization was my second fantasy. However, I never was able to outline the book according to a thematic organization. I found that approach excruciating.

Instead, one day two years ago, I sat down and just started writing. I wrote something that began, "This book is the product of unplanned ethnographic research." The nature of what I wrote was between a journal entry and thinking out loud. That piece is now called "An Unexpected Ethnography." It is part of the second chapter of the book. I thought, "I'll keep going in this vein, retracing the research." I let the book follow itself. It felt unlike any other experience I have had. I wasn't, in advance, directing the writing. It seemed to be happening. After three or four chapters, I laid out a possible structure for the book. Doing that reassured me. I then wrote my book out from the middle. It was thick life-history description.

I wrote most of the book in the first person, although the analytical chapters are more in the third person. The story chapters are in the first person. I am not omnipresent and I am not dominant in the book, but it's like a fictional first-person narrator. Initially, I was not self-conscious about my use of the first person. I am thinking of the first piece I wrote. After that, I noticed that was the way I was writing, and it felt right to me. I think I felt it had honesty to it. It conveyed that I was there and constructing these stories in some way. It also felt like an effective strategy. Then it became a point of struggle with a close colleague, my editor, and some others. I could have written the study differently, but I wanted to write it in a first-person style. I could make up reasons for my use of a personal voice, but I think the reasons followed the practice more than preceded it. I think I felt a strong emotional attachment to doing it this way. I am not sure why. The reasons I can make up are not adequate. I read anthropology. Later, I read the new ethnography. But I came to my approach, and my thinking, independently, before I read any of that.

I feel that, in this book, I am not telling other people's stories. I am telling my take on their stories. I am not changing their stories actively. I am trying to faithfully record, formulate, and interpret stories that have been told to me. However, I fully believe that the stories that people tell me, and that I hear and seek out and gather, are absolutely dependent on my own sets of issues and interests. They are dependent on my own experience, intellect, and background, and everything else about me. I really believe that. Since that's true, why not use the first person, which owns that more directly? It's more honest. I guess it's an antipositivist stance. I do not believe that what I am saying is depicting a reality out there in some pristine universe that I have discovered and am reporting on.

In using the first person, I did not want to be the overt, central subject of my book. I am the central subject in the sense that it's my questions and preoccupations that construct and write the book. However, I am not a white, working-class person, a born-again Christian, a lesbian, or a construction worker. I see these people as the primary subject of my book, although the subject is their lives through my lens. If anything, I have pushed myself to put a little more of the first person into the book, to be a little more revealing of myself. For example, there is a reference to when my household fell apart that is coincident with certain things happening in my research. I have inserted this now in a controlled way in the chapter titled "An Unexpected Ethnography." The paragraph on my household was a hard one to write. It was hard because of my concern about what would appear in print. What impact would it have on my own set of relationships? It is one of the few openly revealing, "confessional" paragraphs in the book.

Only a few other personal-life descriptions are out there in the book. Not much else occurred to me to put in. In the writing, I did not feel that my stories belonged explicitly in the text. I did feel that my presence belonged there very much. I did not see myself as a contaminant in my study. I just did not see the book as about me.

What a close colleague and others objected to was the whole thing: my presence and the details about both my life and other people's lives. For example, I have a chapter on an evangelical ministry that starts with how I felt during a prayer that was said for me. In this prayer, people prayed for my book. They wished their blessing on the Jewish nation, and they wished that my book would be successful and that God would help me to write it and to make it a good one. I start this chapter with the prayer, and then I have a page or so where I write about my response to the prayer, as a way of entry into the chapter. I use that kind of mechanism at various points in the book.

The people who object to the personal material are not especially articulate about their objections. They just don't like it. I think they are embarrassed by it. They think it is unseemly. They think it is inappropriate. It is not social science. It's not academic and authoritative. It's too confessional. It's self-indulgent, narcissistic, unnecessary, and superfluous. That is definitely what they convey. A close colleague says that it makes him personally uncomfortable and that it offends his political instincts. I think he is uncomfortable speaking about emotions in general. However, he goes further than that, and I have had this happen occasionally with others: he takes himself as the universal reader. He says I should keep discussion of my personal responses to a minimum. He ties that to a second criticism of the manuscript: it's too descriptive and anecdotal, too micro. It is not analytical enough. It is not sociological enough. I am expecting too much of my readers.

When people say these words, what do I feel? It depends on who is saying it, and what I think about the person. I do worry. However, I now feel the book is about something. The critics raise the issue of the balance between the stories and the more analytical statements. For me, the current biggest issue is the thick description. I am committed to the personal voice. For a while, I was almost missionary about it. I felt, "Every good author should . . ." Now I think this is one way. This is where I am right now, this voice. I do feel everyone should be honest about who they are, and they should make that clear in relation to their texts. I think you should never take an omniscient voice that pretends you had nothing to do with what you are saying. But whether you use a particular kind of first person as a narrative strategy throughout in a book is an open question. Whether there should be more or less of me in this book is something I am open to. For example, I think about how much, if any, of the voice should be in the introduction and the conclusion. The question of how much of the first person should be in the analytical chapters also remains for me.

When I reread the manuscript now, I have aesthetic questions, not epistemological questions. I plan to write an epilogue that will be an update on the lives of the people in the book, and possibly their reactions to the manuscript, and my reactions to their reactions.

I have said to my editor, "This is what it is. This is the book." There is one chapter that is more experimental than the others. I don't think of this book as that unusual. The experimental chapter is the last chapter in the section about the first family. The end of it, about ten pages, is a selection of summaries and excerpts from my field notes about the main character's reflections and changes. Instead of writing it in a narrative fashion, I did a series as if they were field-note entries. It is not actual field notes in raw form. That would be boring. I shaped these notes to depict the situation and to reflect on the future. One of the scenes I describe in this chapter is when my main character and I go together for a beauty make-over at a fancy salon, at her urging. The chapter focuses, in the end, on my main character's thoughts and feelings, especially about her religious involvement. In closing, I also have some commentary of my own on the textual construction of the chapter.

Eleanor: "Developing the I"

My dissertation was an attempt to make a historiographic argument. It was not a personal document at all. When I got to the end of it, I felt it was a competent enough piece of work. But it did not feel right to me. It wasn't just a technical piece of work—I cared about the issues—but it did not feel like my work. It felt to me like a roadblock that had been crossed. Soon after completing the dissertation, I wrote a summary to give at a professional convention, and I gave it to a close colleague to read. He handed it back to me and said, "I don't think you know what you want to say yet."

Later, when I sat down to revise the dissertation so it could be published as a book, it became a very personal document for me. My problems in revising it centered on how to write the first chapter, the introduction. These were questions of how to set up the book. I had experienced a shift from history to women's studies in my view of what I wanted to do with the book, and that made it easier for me to talk in a personal voice as I started to revise. I wrote the introduction in the first person, dealing with the phenomenon of cultural feminism. When I wrote the introduction, I was elated. The first person was a voice I was happy with. It felt comfortable to me. For the first time, the work jelled for me. Then I rewrote the body of the text to accompany the introduction. I wrote a conclusion that had a moral. It was more like telling a story than the student's exercise of writing a dissertation.

I gave the book manuscript to half a dozen people to read. These were people within women's studies. It worked for them too. Then I started to give it to people outside women's studies, and immediately I had problems. One editor said it was only of interest to feminist scholars. I felt annoyed by that. Then a reader for the publisher who accepted the book—who, I think, was also a man—said it was polemic and only about feminism. These people felt the book was not written for them.

As a result of their comments, I was worried about a possible dismissal of my work. So I went back to the introduction and highlighted every time I used the term "I" or "we," just to bring it to my consciousness: what was I saying when I used these words? At the same time, I was reading some critiques of feminism by ethnic minority women, and I was reading a series of books about intercultural relations among women. I picked up one of the intercultural books one day, a book by a white woman, and there was something in her writing that hit me. She was saying, "As a white woman, I think this." And then she would turn to her audience and say, "And we should do this." I thought she was assuming that her audience was also white cultural feminists. She did this in a way that read to me "exclusion." It made me real conscious of the difficulties of using the terms "I" and "we."

So then I went back to my manuscript to see what I had done. I saw how I had made a whole series of assumptions about who cultural feminists were and what my readership was, and I started to make changes. I limited my use of the term "I" to "I as the writer," and I took out all the "we's" that could be interpreted as anything other than "generic reader": male, female, white, black, whatever. Because I got so worried about the issue of exclusion, and because I changed my uses of "I" and "we," the final draft of the book is different. It's not impersonal, but the extent to which I use the terms "I" and "we" is much less, and I have tried to be very careful about how and when I use those terms.

The result is a draft that is still in the first person: my voice is in it as a person, and I'm trying to address the reader in a personal way, as "we." But I try not to make assumptions about who the reader is other than that the reader is someone who is following my story. Although men had initially brought my use of "we" to my attention, by responding to it as exclusive, it was easier for me to see it as a problem when the issue was raised about exclusion of ethnic minority women. Ethnic minority women are the center of the work, and that exclusion was both more hidden from me and more troubling. The response of the men mostly made me mad. Considering the ethnic minority women's issue made me want to change things. Now I hope the book reads a little less exclusively. It also puts a greater distance between me and feminist scholars, and I feel that is a problem with it. The book does not read to me as though it is a conversation within feminism now, as it did at first. It reads reflectively: here is a particular person reflecting about a particular subject and trying to draw the reader in with her. But my fear is that, as a consequence of taking out all those "I's" and "we's," the book reads more as a critique of feminism than it did at first. It reads more like an outsider's view now. So there is that cost.

I originally chose to write using the "I" in order to get my work published as a book. I had read someone who talked about a book as a humanist statement. But what made the biggest difference to me was that, when I sat down to write the introduction, I felt I wanted to put the book in the context of feminism. I then read cultural feminist literature extensively, which is contemporary feminist work, and different from historical work, and it affected me very personally. It felt personal to me to read it. It also gave me permission to talk personally in my own writing in a way that was hard to find within history. Historians do not talk in the first person very often. After I read the cultural feminist literature, I sat down to write my introduction, expecting it to be a miserable experience, and it wrote very comfortably and almost immediately. I think I would have written the introduction in the first person even if I had not read cultural feminism. But I probably would have written, "I am interested in this subject because . . . ," and I am not sure that I would have been able to maintain the first person, to continue to use it. I was affected by being immersed in material where it was okay to use the first person, and where the women who were writing felt a women's studies obligation to locate themselves individually.

In the first draft of the introduction, the "I" was legitimation for me. I was saying, "I am a cultural feminist. I am in a mainstream." I felt very good about that draft. It was a quantum leap in doing the work. Ever since then I have felt comfortable with the manuscript, and I feel the changes I later made with the "I's" and "we's" have more to do with how to approach an audience than with changing fundamentally what I wanted to do in the book. The initial uses of "I" and "we" made the work feel like it was mine. On the other hand, I have growing doubts about how wise it is to self-identify in other contexts than writing. Some of the issues are similar, so I would like to mention them.

In my teaching, for example, I feel an obligation, as a women's historian, to identify myself in personal ways to my students. However, if I say to my students, "I am a white lesbian feminist," I am aware it does not work well. All of the students who cannot fit themselves within any of those categories immediately dismiss what I have to say. Or the opposite occurs: those who do fit themselves into the categories then begin to learn from me through a process of identification that scares me, because it is not very reflective. For example, lesbian students who identify me as a lesbian faculty member may feel real good about that, but they then tend to assume that I will like anything they have to say. They assume I will agree with anything they say and that we will all get along just fine. They assume that I do not really mean what I say about critical writing or critical thinking. So I end up having mixed feelings about self-identifying when I teach.

This difficulty does not arise for me so much in writing. When I read other people's writing that does not have a first-person voice, I find it stilted. I think I want the writing to be a different form. I want it to be a conversation. Not a lot of academic writing is conversation. So it seems to me that to put my own voice in it, to make it a conversation, is a straightforwardly positive thing to do. I am speaking of doing this within the limits of what I have been describing about how I say "I" or "we." I think my first-person voice makes my written work more accessible. I guess in teaching I don't worry so much about accessibility because I am already having a conversation directly with the students. In both the writing and teaching, when I limit my "I" and my "we," I feel that reflects an attempt not to be dismissed by others on the basis of a self-identification I might make. There are two issues involved: When do you use "I"? And what set of code terms or characteristics do you attach to that "I"?

In the final draft of my book, I think I come across as "the writer of this book" and not much else. In the process of writing the book, I took out some of the limits on who I was, and I took out a lot of the limits on who the readership was. The whole thing felt to me like a process of learning. Developing the "I" was part of developing a voice and learning to speak about what I wanted to discuss. I found that I had connected my "I" to so many things at first. There was a whole series of identifications of who I was. However, in this last draft, many of those self-identifications are gone, because putting them in the draft made me think about what they said about me. It made me think about who I was, and it made me consider issues differently than I might have otherwise. I concluded that it is not quite so simple as, "We are the products of whatever our background is." That was not how I wanted to present myself.

In the last draft, I do not identify myself with the label "cultural feminist" as I did at first. I say, "I am somebody who . . ." In the end, it makes the "I" more personal, because it is less coded, less attached to code words like lesbian, feminist, scholar, white woman. Within contemporary feminist writing, however, the tendency is to do the opposite. The trend is to say, "I am a white woman writing about ethnic minority women. Therefore, I have a whole series of built-in biases and prejudices and everything I say has to do with my background." I feel there is a great deal of pressure to identify yourself in that way.

I think my use of my own first-person voice is still developing. Since drafting the introduction to the book, I have found that I write things more easily in the first person. This includes conference papers, discussions, even lectures. I write them more easily than when I try to write in a more scholarly style. I am thinking of a conference paper I wrote that worked very well, that, at first, I desperately did not want to write. Then I sat down and I wrote it in about five hours. Use of the first person gave me the freedom to talk speculatively. I think that paper has weaknesses; it is not comprehensive, but it worked well in the setting. It came across as reflective conclusions of an individual person. People responded to it in a way that was very positive and thoughtful, rather than saying, "This is good or bad work."

I am trying to recapture that approach in a paper I am writing now. I am toying with the idea of writing the paper both ways. At first, I thought it would be easier to write it in the third person because I was so steeped in the literature. I thought I could easily write a generic, third-person review essay. But I have started to write it in the first person instead. When I began to write, that's what happened. The difference is, if I were to write a generic review essay, I would feel an obligation to cover everything. In the first person, I can say, "This is what I think is interesting." Whenever I do that, when I use the first person, it's faster. I stop worrying about footnotes. It feels more like a conversation to me than a term paper. Self-censoring occurs for me more often in the third person. I think that for women, in general, and me, in particular, the temptation to self-censor is often overwhelming.

My feeling is that I would be better off to be more playful and more creative in my writing. But these qualities are not what I experience myself as particularly good at. What I know I am good at doing is reading lots of material, making sense of it, and figuring out the themes. What work I have been able to do comes out of habits of excessive organization. You go step by step and eventually you get to the end of it, and you make some sense of it. The process of putting a voice into it, of making it creative, is something I experience as harder. I think a lot of women experience it as harder. I think we would be better off if we did not. I would like to be able to take more risks than I sometimes feel I can. For me, it is a learning experience to try a different way.

I think it is particularly hard for historians to use the "I" in a creative fashion. I don't think people go into history because they are creative. Historians do not tend to use the "I." We teach ourselves not to. However, great historians (many of whom are men) do use the "I." They feel comfortable speaking in the first person. To the extent that feminist studies allows women to develop the "I," then it becomes almost an alternate route. Nonetheless, the situation will affect when I use it. Orally, for instance, I don't use the first person very much when I talk to men. This is because they use it to dismiss you, in departmental meetings, for instance.

Kathleen: "I spend hours trying to squeeze it into the formula, when I could be spending hours observing children."

I do see an order in the way my investigations and my writing up of them have changed over the years. Having come from a background as an English major interested in creative writing, my first attempts to observe situations and report on them were in the style of Tillie Olsen. My first major attempt was to write about a day-care center where I worked. I took the first-person point of view intentionally and wrote down my observations, impressions, and descriptions of the interactions among the children and parents. As I reread that manuscript now, it is clearly full of me. It is almost embarrassing for me to read it now. I find I want to treat it as a piece of data. It is embarrassing for me to read because I talk about my feelings and opinions. There is a certain passion in the writing, a conviction, and a missionary spirit that now I am constrained to call rhetoric and to see as misplaced subjectivism. The embarrassment I feel comes from the Ph.D. person I am now. If someone else had written that manuscript, I would say, "How refreshing." However, now I am trying to figure out how I can use it in an academic sense. One suggestion has been to frame it in terms of a set of theoretical questions and then to look at everything I said as data.

I wrote that manuscript before I went back to graduate school for my master's and Ph.D. The day-care center study was a transition piece of work for me, from creative writing, like fiction and poetry, to writing about real life. When I started in graduate school, while working on my master's, I remember being scolded for bringing in a personal note about my daughter. In a paper I wrote, I described her looking out and seeing patterns in the branches of a tree. I thought it was a good illustration for the paper, but I was reminded that you do not use personal-life examples, even in an essay for a class. In my Ph.D. program, this was even more pronounced: not to call on these homey little examples, not to use my personal experience of life, or of my home setting, and definitely not to say what I believed. I found that my writing was constrained by professors who would very often say, "This paper is fine up until the last two or three pages." In the last two or three pages was where I would sit back and say what I made of it all. I would give my interpretation. Not a reiteration. I wanted to say to them, "The only reason I did all this was to tell you what I think of it and what it provokes in me: what I can conjure up, or imagine, or predict." What I wrote was often called rhetorical, opinionated, or personal. So I, in fact, then felt bad about putting those things in again.

A similar thing happened with my attempts to present a feminist interpretation. When I used words like "patriarchal," I was told by people very early on, "That is a red-flag word. It sets people off. They stop listening to what you say if you use words like that." "Patriarchy," "sexism," all those words, were unacceptable. What I found, as I was trying to develop some academic or scholarly voice during my years in graduate school, was that the theory I was learning was male theory. Nonetheless, I wanted to use it to show how the logic of the sociological theories themselves produced important contradictions, particularly when it came to the treatment of women and children. It was a rather tame thing to do. It was not like I was out to destroy all male theory. I was using the theory. I never understood why, when I was applying these long-accepted theories to women and children, it suddenly became rhetoric.

My writing style changed very drastically in graduate school. It became very stilted. I started reading dissertations to figure out how people wrote them, not the subject matter, but the style. I was mimicking that. I think I soon got trapped in it. I found that it was acceptable to the publishing world, to whatever those powers are: the people who are hiring for a job, or looking at a paper for a journal. There is a style that I would define as very distant, clinical, full of jargon. The words and forms it uses are the opposite of red flags. These words and structures, when the people who read them see them, are comforting. They are very comforting because they are key words that make people say, in effect, "You are a member of our club." I think there is a de-sexing of the author through this language. I am not happy with that. There is clearly a formula where you start out a paper saying, "Many studies have been done but this is where the hole is." Then you provide some theory, then data, conclusions, and suggestions for future research.

The formula is very comforting, but, to me, it is constraining. I do not let myself spin a thought out in that formula, or really do intellectual work. I want to say what I think and believe, what my attempt is at making sense of what I have read, not, "At the .051 level, this shows significant correlation. Therefore, I conclude . . ." My wish is to say, "This is what I think about something," and then the implicit question to the reader is, "What do you think about it?"

Recently, I have been looking for a forum where I can say what I think, based on my experience, on reading, and on research in the field. But I think it is almost a matter of trust— that I want to be trusted by the reader as someone who would not sit down to write unless I had studied my subject enough to really have something to say.

It is confusing. The formula does work. In a paper I am working on now, the comments I have gotten back are methodological. I use material that is anecdotal. The reviewer I am having to answer accuses me of choosing quotes from books that prove my point. He says that I am selecting material. I think what I am trying to do is to throw an idea out there, to say, "This is why I have come to believe this phenomenon is important for us to look at." I really feel caught. In graduate school, I felt caught between wanting to pursue the cross-national sociological research that was almost totally quantitative, that I was involved in, and pursuing simultaneously a feminist viewpoint. Now, I am freer. I am able to write about things without the quantitative core. But I am feeling on slightly shakey ground because I am leaving parts of the formula behind in order to develop my own voice. As I sit here trying to do my revision, I hear my graduate adviser saying, "It is too broad." But then I think, "Of course it's broad. I want to have the larger discussion."

I was talking to a colleague of mine. She said that, as she sat trying to write, all she could hear were the critical voices. It made her absolutely block up. She had to extricate herself from the critics' voices first. I, too, picture those men in my mind. I see them wagging their heads back and forth. They are saying, "You don't. You have got to . . ." They are concerned with the way I use the concept of the state. It can keep you from saying what you want.

In work like I am doing now, I don't see any place for my personal voice, except to say, "In this paper, I will argue . . ." I am de-sexed, neuterized, unmotherized, ageless, I have no personal characteristics whatsoever in this paper, except for a Western bias. The insights that I have as a childcare worker, and as a parent, have no place in it, and that upsets me, because that is something I want to be drawing on.

The formula style is not the way I want to write. However, acceptable journals require it. And, apparently, that is what I need to do to stay in this game. I am concerned about tenure, but I don't think getting tenure is the only pressure. Around me, people, if they have tenure, suffer if they do not write in this way, using the formula. They suffer disdain from colleagues. There is little outlet in my field for alternate ways of writing. Even in feminist scholarship, even *Gender and Society* and *Signs* follow basically the same formula. In education, if I were to write for a more applied journal, I might get something different published, but whether or not I could put it on my vita is not clear. Although that may be my scholarly snobbishness. I think there are respectable journals and I am not taken seriously unless I publish there. In the field of education, there are publications for practitioners. There is an attempt to rethink the theory and practice combination repeatedly. But the more you fall on the theory side, the more seriously you are taken.

What about emotions? They are so much forbidden that I don't think I ask myself how I feel. Outrage is one thing I feel: outrage that the system is unfair. What is more forbidden is to say, "What the world needs is love." I might believe, in my personal life, that the attitudes we have toward each other are far more important than the scientific conclusions we can draw. But if I were to propose that in a paper, it would be tossed out right away. "Go talk to the human growth people," they would say. I do not see any place for my emotions. I think that the more hidden they are, the more credible my writing will be. I am embued with the teaching that you cannot learn anything from an *n* of one. The way you can include emotions is to do a survey about emotions, other people's emotions, and say, "This many people felt this way. This many people felt that way." Then you could do a theoretical introduction about the sociology of emotions and write a conclusion.

How do I feel about this? I feel I have swallowed an awful lot of the male, positivistic, scientific point of view. To the extent that the personal and feeling point of view is a female point of view, it makes me understand, again and again, that the whole field that I am in is not only male-dominated, it is thoroughly male-defined. To the extent that I want to write in it, I must become male. If I want to be a female in the sense of expressing the personal point of view and showing emotions, then I have to find another arena. It is not here in the university. My medium for writing my feelings has always been letters.

What if I was able to express my emotions in the academic realm as well? What other emotions might come out? There are variations of anger, including frustration at the very formula you have to follow. What I feel in writing social science is a frustration, a desperation. I am wanting to do it right. But there is a lack of satisfaction in being able to truly say what I want to say, and say it clearly. I feel frustrated, desperate, out of place. There is a way I feel I do not belong to the club that writes in this way. About my subject, the societal care of children in the world, I feel outrage, and an outpouring of caring, and a desire to put that into action. But in the process of trying to put it into the formula, I sometimes feel disgust. I hate that process. I hate what I have tried to squeeze my material into. It no longer says clearly what needs to be said about my subject. I never want to read back on it, even when I have to do revisions later on. At some point, I realize that I am spending hours trying to squeeze what matters to me into the formula, when I could be spending hours observing children. That is where the disgust comes in. I have spent so much time trying to fit the form into the dress that I no longer have any feeling for the body. This image I am having now is of stays and girdles that you pull to make it tighter. What do you do then to the content?

When I write letters, I use it almost as a forum for saying my thoughts and discussing how things are going. It is like what we used to call an essay. There are no footnotes. If I could have a job where I could express things in a journal or a letter, I would be delighted. In graduate school, when I would feel charged up by a class or a lecture, combined with readings I had been doing, I would sit at a typewriter and write out how I thought these things applied to the problems I was thinking about. I was integrating the various sources. That is really what I think I can do in journals and letters. These are free-form notes: notes to myself. In these, you can integrate. You integrate, into your own whole knowledge, your experiences, and what you have read and heard in talks and lectures. And you can express the integrated whole. Whereas in the scientific writing, I find that I must break it back down. I must say, this is the information that I got from reading, this is from observations, this is from collecting data. In the scientific writing, there is no place for the personal material. So you get a less integrated product out of me.

Right now, the two main outlets for my integrated writing are letters and my own journal. Those two areas are where I can take how I am feeling that day and relate it, or integrate it, into what I have learned, what I have observed, the data I have collected. I am allowed to do that in those media. I am allowed, by the rules, to give it all to you, to give you all of the sides of it. It makes scientific writing very meager by comparison, very poor, very sparse. In other words, we deprive each other, as readers, of this integrated view, because of the way we have come to think, and because of these formulas we have had to use. Writing letters, or writing in one of these integrated forms, might be a good way for me to think about how to use my day-care material. I should not see it as a detriment that the manuscript was written as it was.

But who would publish it, if I dealt with my experience in a different way than treating it as just data? Who is interested in that? Who cares about that? The first thing for me, I think, is to become more explicit about my methodology. What I have used, thus far, is not a methodology that I have chosen. It is one that was given to me, along with my academic hood. So it is important for me to start thinking about an alternative. However, I am not really interested in methodology, except that I know I have to become more conscious of it. I don't tend to think, "I want to do qualitative work." I think, "I would like to write more of what I believe." I have seen the people who do qualitative work. They have to defend it up and down the wall. I don't want to do that. No one who is writing a quantitative dissertation has to give any excuse for it.

I have been thinking in terms of articles, but there is the book I am supposed to be working on this year. If I could allow myself to sit down and start telling this integrated story, I would write quite a different book, and it would be far more useful, I think, and certainly more readable than the articles are, if I would dare to do it.

Diana: "I am truth."

When I want to write something my way, I have felt there is something wrong with me, with my style of thinking. In my graduate work, my academic writing would be well done. I did well on exams. I got good responses. I wrote poetry for assignments. I did a paper on existentialism that was a play. Recently I wrote a paper for a cognitive psychology course, which was required for my psychologist's exam. The paper was really good because I never once did what they asked to be done. However, I did stay within some limits that would make it acceptable to the authority. I feel I almost cannot do something the way the authorities want it done. It goes so against the grain of me.

In my dissertation, I had a struggle about whether my way had value. I feel that my way just does not go in the mainstream. There is a line from Eliot, "That's not it at all," which seems to me the struggle I have all the time. I come up with something and someone is saying, "That's not it at all. That's not what I meant at all." It is my experience that usually I have said something that is so far ahead of where they are that they cannot track it. They say, "That's not it." I believe them that I am wrong, and I feel humiliated. I wish I had the courage, instead, to push my beliefs.

Just to do a qualitative dissertation was deviant in my department. I initially spent one year writing a chapter on the use of imagery among battering couples, seeking to measure it with statistics, because one member of my committee could not grasp what I wanted to do as a qualitative dissertation. I spent a year writing the chapter, got it approved, and then dropped it. Meanwhile, I became fascinated with anorexics as they came walking into my office. However, the literature on anorexia did not fascinate me. I then read the object relations literature. I saw parallels that did fascinate me. I began to look for a way to write a dissertation that would explain both to me and to the girls, and to an audience, why they saw the world, and themselves in it, the way they did.

I found a committee for the topic who were social scientists, not psychologists. I had people who were willing to look at it my way. In writing my dissertation, one of the hardest things for me to do was to get a basic structure. I followed a formula. I copied other dissertations to make my work fit the system. Within the confines of that, there was a large section in which I felt I was free. At the time, I was terrified of my writing. I was afraid that I could not do it right. When I write, I write and write and I do not know what is pulling it together until the end. After spending two years writing the dissertation, I spent one week writing the conclusions, which were the most interesting and important part. The conclusions did not take me much time because the study had finally come together. When I think about that experience, I think of the creative process and the fact that you have a problem. You have to do a lot of research and you struggle and you struggle. I struggled through the entire two years not knowing why I was doing it, and what the direction was, until the end.

I have always wanted to make the product of my writing be a demonstration of what I am trying to talk about, rather than it being just words "about" something. I am constantly thinking in terms of an art project in which, when you open up the book, you do not have to read the words to understand it. It is what it is talking about. Just talking about something does not seem sufficient to me. My dissertation, as a whole, was not a clear demonstration in this way, however, because there were too many things I was required to do. But, within it, in the case studies about the girls, there was that kind of layering, of words on top of words, so that you could begin to get a picture of each personality. I think that is a significant problem for me: how do I manifest "what is" when words are not sufficient?

A paper I wrote on interpersonal boundaries was also an example of a study where I had no idea what I was going to get. I just suffered with it and, all of a sudden, it came together. That paper did not follow what was asked of me either. I feel that I break the rules, but I do it with enough interest, or excitement, that I am forgiven. I don't break the rules grossly. I am not that kind of iconoclast. I do it just subtly enough so that I do not get punished, but I maintain my own integrity. In both my dissertation and my boundaries paper, I was taking people's words and trying to understand their intent, or their deeper meaning. I took the words verbatim from therapy sessions. I felt the meanings of the words were embedded in each particular personality and in a variety of contradictory words. In both studies, I was trying to understand why my clients would say certain things that were often foreign to me and that I did not understand.

I think people are attempting to communicate what they think is true, but they do not know themselves very well. I was looking for how the anorexics, and others of my clients, were attempting to protect themselves and to communicate. The process was one in which I had to believe their words. Then, in a layering way, I looked at the different levels of interpretation. I would also ask them. I went back and asked my clients, "Could this mean . . . ?" I was attempting to get an internal logic for what everyone was saying and doing. At the same time, I knew that, according to the world and its rules, every one of my clients was sick or crazy. I did not think any of them were. I knew we were all struggling, but I felt we were internally coherent and logical. Every step we took, and every word we said, had some strong positive intention.

In trying to understand my clients, I think I was looking for the same thing I experience in my life. I feel there is this big system that lays over us a certain way to be, and none of us can manage it very well, me nor my clients. I believed that my clients had a logic and that their idiosyncracies came from a deep, internal logical space. That was what I was looking for. I feel that those of us who keep breaking all these rules—to be creative, to say, "This is me"—if we have struggles with it, we are also trying to say, "You know, I am really part of this system too. I want you to accept me, but not at the cost of loss of my integrity, or loss of my sense of self." I think my clients are saying that too. There is a coherency to each one of us, but we are not understood and we are hurt and we're angry.

Writing up my research, I thought about my clients and about myself. I followed the method of compare and contrast.

Identify. But although I identified up to a point, at some point, I did not understand, because if it were me, I would not do, or think, some of the things my clients did. At that point, I would have to say to myself, and I used "me" every time, "Well, if I did that, what would it mean? Why would I do it in this strange way that this person did do it? Or in this way that seems strange to me?" Then I would have to go back to me. I feel I am always the reference in interpreting anyone, even if I am using their words and even when I do not understand them.

To comprehend, I have to use me as the reference. If something is foreign to me, I could just write it down, but then it would no longer be me. It would simply be a report, with no life in it. Maybe that is what I hate about a lot of the things they want us to do in academic writing. I feel it has to have something of me in it for it to be alive. That is why I can't stand psychology journals. The articles are just a report. Whereas I am very interested in what people think and feel, and in how I relate to it.

I keep coming back to the rules of the world. They seem to me to be laid on. They do not seem to be an integral part. They are not the coherence in all of us—the operating principle that is true to each self. The problem I see is how to maintain a regard, or respect, for the operating rules of the person. It could be a paper in a class. How does the professor grade it and still maintain respect for the rules of the person?

I think that the purpose of all the larger system rules, whether they are the rules of the university, or a publishing house, or a journal, is so there is a simple manifestation that has continuity. There seems to be a general belief that we would have anarchy if we did not follow all these rules. I don't really believe that. Even chaos has organization. The mystery is, Where is the organization? I think that is what everybody is doing. They are looking for it.

Within complexity, each one of us wants to grab hold and say, "This is my version." Each one of us is manifesting some kind of truth, or statement, about existence, "isness." We each manifest our statement at a little bitty level, like in how we do something that everyone else does, like cook carrots. We try to do it in some little way. Then altogether, in a myriad of ways, we make a larger statement. Everything that is happening is a statement. It is that layering idea again. Whatever one does, if it is done in a way that is real, and that has a sense of presence, then it is a manifestation of "I am what I am talking about. I am what I am doing. I am what I am writing. It is not about it. It's in it. I am not talking about truth. I am truth. But so are you."

The idea that there are these rules of the world that we should all follow is just the opposite of what I am saying. My manifestation of truth is just slightly different than yours. If you tell me to manifest truth your way, then anything I do is a lie. That is the struggle I have been having all these years. I feel I keep running into somebody else telling me to manifest truth according to their definition, and I can feel the lie of it. I refuse to do it their way, to lie, because I can't. Then I live with this terrible fear of being punished because my truth is not their truth.

Lots of us get punished because somebody out there has made a decision that their truth is the truth. That is why I keep going back to art. It is the most direct form of truth each of us has, whether it is music or dance or decorating a home. I feel that if I can really express my truth and merge with my work, as an artist, if I can lose myself in my own personal production, I become more like you. I become more like what is deep and alike about all of us. It is paradoxical. As I attempt to be more myself, I become more like everyone else. I become more like you, but not like the rules of the world.

When I think of the clinical psychology field in which I have been trained, and the guidelines for understanding people in my field, I feel the anxiety that I am not good enough. I compare myself to those people in the field who behave according to the way the field is presented, the bright young men. There are bright young women, too, but it always has that feeling for me of the bright young men, who know a lot about the field. They have memorized many facts. They understand statistics. They can quote the DSM III. They usually wear ties, or, at one time, beaded necklaces and long hair. No matter who they were, or whatever costume they wore, I was out of synch. I wanted to learn all the material they knew. I wanted to wrap my mind around it, but I could not. It was too boring.

I think there is some deep resistance in me to what they

stood for. I think, "What's that got to do with helping someone through their suffering?" The whole field of psychology and counseling has problems like that. I see the journals. I like to read some of the findings, but the research is all about counting behaviors and trying to make it a science. To me, the research is valuable. It can be interesting. I think in terms of what it could possibly mean theoretically. However, I think it is somehow destructive to the people who do it because it is stultifying.

Gender makes a difference. Some research indicates that in a test they give for high Machiavellian personalities, the high Machs predominate among M.B.A.s and psychologists. These people are out for their own goals, not for the goal of helping others. I think there are very few liberated males. There are many who want to be, but they do not want to do what it takes. The field of psychology is dominated by males, particularly where it is "important": in the administrative part of the field, and in the faculty of universities. I think males are socialized to be condescending to females and to think they are better. In psychology, men look at their findings more as "the truth." Women are more likely to be more uncertain, to suggest, "This is one way to view it," and not to put limits on others. A male will put his limits on you. Females do not do that as much.

In all my schooling, I have felt that I do not think exactly right. I do not think in a sedate way. I feel as if I am going to be in trouble any minute. It scares me. Yet I can only think of something the way I think of it. I do not really know how to accommodate. This came up recently concerning publishing. Somebody else has to teach me how to accommodate, how to do it right. An example that also comes to mind is when I tried to get into graduate school for my Ph.D. A friend of mine took me to lunch with the faculty member in charge of admissions because I had blown it with him when I went in for my interview. He had given me a catch-22. "How can you get into the Ph.D. program if you have never done an internship?" he asked me in the interview. I said, "Well, I'll get in and then I'll do an internship." He said, "You can't do that because you'll already be in, and you would . . . You can't." Then the tears started coming down. He said I decompensated. My friend took me to lunch with him a few weeks later. I never said a word at lunch. I felt I could not talk to this man. My friend, a woman, just sat there being Southern and made it all work out. At the end, this faculty member said that my papers had risen to the top, and I got in.

What I am trying to say with this is that I feel I am always in trouble. This man represented the establishment in the old crummy meaning of the term. He represented all the blocks, and the fact that there is no way I can take care of myself with the establishment. I feel that somebody else has to do it for me and I better be quiet. I do not talk in classes when I am a student. My experience is that nobody understands me, or they correct me. Then someone else says the exact same thing a few days later. It pisses me off when that happens. I have wondered if maybe there is some unique way that I present that causes me problems. On paper, things are not so bad because then someone can take time to look at what I am saying. Their immediate reaction may be, "What the hell is this?" But then, somehow, I get a positive reaction.

I feel I cannot give it up. I will not give it up. Whether it is talking to people, or not, or writing something, or putting on a program for stage, or teaching, I refuse to do what I basically know I have to do to get along with people, because I feel that would be total death to me. Whatever I do, I feel that following the external rules is death of the self. I believe the cliché that one must die this moment in order to live the next. However, it is important to me that I make the choice, by my internal rules of life and death, not somebody else's. I have been extremely aware of the question of who makes the choice, and of the issue of life and death, because of working with the anorexics. Part of the fascination I had initially was some kind of identity I felt with them, and some incredible grieving, both for me and for them, that we each had to give up something of ourselves to survive, but the anorexics to a terrible degree.

Do I use the first-person singular in my academic writing? No, I would not want to do it. I do not really like it. I am not interested. But embedded in the words I write is me shining through. The "I" comes through. I choose not to use the word "I" because I think I am too exposed when I use it. However, I think I am manifest in the material, even when I speak in the passive voice. I might say, for instance, "There is a great joy in . . ." I am aware that in other people's writings, I do not object to the "I" when there is not so much of it that I am inundated. When an "I" is used that I do not find intrusive, I hear that the author is embedded in her material. The use of the "I" is not the important thing for me. It is the idea about the "I" that I like. I am interested in the content. The content is the writer somehow. I feel one can use the word "I," but you do not have to. There can still be presence. When I read material with the "I," I get these images and pictures from it. The person who wrote it is part of them. It is like a mosaic, where the whole is made up of many little tiles.

15

Problems of Self
and Form, II

Ginger: "I think about play writing and social science"

FOR MY MASTER'S THESIS, I felt strongly that I wanted to write in the first person.[1] My chair said, "Keep it to a minimum." I tried to use it when I really felt the "I" was there, such as in the first chapter concerning why I wanted to do my study. My dissertation was more a trial because they like to put you through it. My committee chair was very negatively critical. However, I also had trouble writing the dissertation because I had so much data. I just did not know what to do. I could have benefited more from working with people who did ethnographic work. I could have had more models. I kept reading other people's studies, thinking, How do you write this stuff up? I think the writing up is difficult because there is not a set formula. You want to do justice to your data. You want to present it in an interesting fashion that does not kill it. You look for a way to bring this wonderful material to life. It is very hard to do that in a social science study.

In my dissertation, I also lacked confidence. With the play writing I am doing now, I feel similarly. You have to figure out your own way of writing it. I have to do it in a way that satisfies me. I believe in an idea of thick description. I think you have to spend a good deal of time in the field. Also important are triangulation, analytical induction, and theoretical grounding. And being a decent writer is important so that you get a sense of vivid description.

I feel satisfaction both in play writing and in social science writing when I get an insight, when I put things together in an interesting fashion that might shed light on something. By "interesting," I mean something that has not been put together before. The satisfaction comes when I feel as if I have made a discovery. I feel excited: "Oh, that's neat. Maybe this was worthwhile after all." It is like an ink blot. You get people to see something they have not seen before. Or, if they have, you have not heard about it. It's a high, a find. I think the interesting thing about ethnography is what you make of it.

In play writing, I will figure out something about my character or story. Things come together. It's an intuitive flash. Ultimately, you see your work on stage and you like it instead of wanting to throw up.

I think about play writing and social science, the similarities and the differences. In novels, you have a narrator, like in social science. But in plays you do not. It is only the characters. I find that very difficult. That is why I like my characters to have monologues in which they speak directly to the audience. I think that is the author saying, "Look, let me tell you what is going on here." When I wrote my most recently produced play, *Woman Seer,* people said, why didn't I put someone like me in the play? I could have put in a person who was studying these churches, or who came in for some other reason. However, I felt I should know the milieu well enough to tell the story from the insider viewpoint. Someone like me would be an outsider. Such a character would be too removed from the spiritualist church situation the play deals with.

I think ethnography takes guts, but play writing takes a lot of guts. You are so bold as to feel that what you have written is something other people should act in, and direct, and other people should come and see it and pay money, and actors should spend six weeks of time on it. You really have to involve a lot of other people, both in the development and the later presentational stages. I like that because writing is lonely. At the same time, you end up having to satisfy others, like on a dissertation committee. I could not just write anything I wanted to in a play. You cannot be totally free.

Am I totally free anywhere? No. But maybe in my journal. I have been keeping a journal since I was eighteen. That is one reason I had the guts to say to my thesis director, "I want to write my thesis in the first person." Because of my journal, I think I am more comfortable writing in the first person than most people would be. I write a lot of my thoughts in my journal, both intellectual thoughts and describing emotions. I am probably looser in the journal than in my other writing. I try to get my thoughts down. It is not nearly the editing process that I do with my other writing. I try to shut down my editor. I try to get my intuitive thoughts and feelings. That is what you really want to start with for any creative writing work. When I took a journal class from Kate Millett, I learned about not worrying about hurting people when I wrote. Women always worry about that.

The function of my journal is to get things off my chest, do-it-yourself therapy, and to get ideas for my writing. I used to keep my journals in notebooks. Now I keep them in file folders because it is hard for me to separate out when I am writing for myself and when I am writing for my different writing projects. So my journal is now all in file folders, each for a different project. Except social science is in a separate folder. My journal ideas do not as often end up in my social science folder as they do in the folders for my creative writing projects.

I see a difference between creative writing and social science writing in that I feel, at this point, that I know better how to do social science writing. I used to enjoy it. However, it is always hard at the time. All writing is. I am torn now because I get a lot of invitations to write social science papers for conferences and books. For me, that is a sure thing. It means I am going to get my name in print and my little points. I am interested in the subject that I write about, gender and communication. But I would much rather now write creatively. It is more challenging. It relates to a deeper core. It is the higher goal. In my creative writing, I tap into emotion more deeply than I can in social science writing.

In social science, even if you use the "I," I think it is not as personal as in creative writing. We are still bound by the convention that you should not be too personal in social science writing. My social science humor article, for example, about how women use humor in conversation by themselves, is one I wrote in a more personal way. However, there is still that distance. Even if you use "I," you are still using a distant "I." It is like the difference between an essay and a poem or a personal short story. Maybe, when you are a social scientist, the first person means, "I was there. But I was not a player." We know there is no such thing as objectivity. The best thing you can do is to be honest about your subjectivity.

In social science, I try to do an emic approach. I want to understand a person's world from the inside out, and to be true to that person's terms. In a play, I am trying to do that too, although you can distort anything you want in a play. With *Woman Seer,* I collected all this material about spiritualists. I did interviews and visited churches. I went to a spiritualist convention, and I took mediumship classes. I wanted to show this world. What if you did believe in this world? It was when I turned my research into a play that I felt it became different from social science. In a play, you have to tell a good story; fiction enters into it right away. At the same time, your story has to be true to the situation, and it has to be true to yourself. Otherwise, it will not have any heart.

It took me a long time to figure out how to put myself into the play. My main character in *Woman Seer* is a fiction based on interviews I did about a real woman medium, and on what I had heard about another woman. I originally thought I could do a play about the real person. Then I realized I did not know enough about her to do it fully based on her, and people were not going to tell me enough, because they revered her. I also did not want to mess with their high priestess, so I thought of a more fictionalized story, and that she would be the inspiration for it. I felt that nonfiction would be tampering too much with my interviewees' little world, even though my play was not going to have any big impact on them. I also felt that by fictionalizing their situation, I could be more honest. This way, I could portray my character the way I wanted to, without worrying about how people would react if they thought she was the real woman.

In addition, there was the time period. I made decisions about the historical period that affected the story. These were decisions that you would not have the option to do in social science, like setting the play in a different time period than the research.

I struggled a lot with that play to find the real story, to move people, and to connect with my emotional center. I struggled to determine what conflicts in my life could relate to the conflicts in the play. The woman I wrote about is a very powerful woman in my story. I think I picked my subject because I am interested in women and power, and in how women usually have power in offbeat things. I have had to see how I relate to my subject because when you write about something that is not connected to you, it is not interesting. It does not have emotional content. If I was just going to write about this subject for social science, I might not have felt I needed to connect with it in a highly personal way. Because then I would have been writing "Mediumship as Communication," or "Spiritualism as Folk Psychology." In social science, you are not supposed to be so engaged, whereas in creative writing it is the other way around. You want to get in there with your material and wrestle around with it. You always are in all your writing, but, in creative writing, I am trying to make those emotional connections that make it a powerful and touching piece of work. I want to touch people, to make you feel something.

Social science can be touching too. But there usually is more of that distance.

What if I took my passion and emotions and channeled them into my social science, like I do in my plays? Does social science have to be distant? My first response is no. And that we, as women, should re-create something different. Because the distanced way is how social science has always been done, and men fucked us up, and it's boring, and they were trying to imitate a model that is not really true: logical positivism. However, I have not figured out what social science would look like if it was different, beyond the modifications we have made so far such as using the first person and doing more qualitative research. I am intrigued by the idea of something different. I don't know why I turned away from social science. I think we are all trained with the traditional forms. To think of new ones, it is like you are being asked, "What if this wasn't an eagle, what would it be?" Then you have to imagine an eagle in a chicken suit. How would it then be social science? Let's say we were going to make a new definition. I think it would have to involve research, and not just my imagination. I think of field research. Systematic research.

When I did my play, I probably did more thorough research than anyone else in my group. I was aware that I was doing a very thorough social science job of it, to satisfy myself.

I had to learn that by fictionalizing, I sometimes could get closer to the truth than by telling the actual story. I had to think, What was the story I could tell that would do justice to what my research indicated? The fact that I had to put my work on stage meant my story had to be more dramatic than social science in order to heighten the issues. I felt that was okay in a play.

As a playwright, I feel I have different options for expression than I do as a social scientist, but I also have to abide by conventions. If you look at the convention of classic dramatic structure, for instance, it looks like a male orgasm. I don't think women write that way, at least not some women who are not inculcated into the patriarchy. Our lives are different than men's, and I think we tend to write stories that are more circular. I thought play writing would have more freedom. But I found that it is this view of how the world is: action and reaction. I think life is more subtle and complex than that. Characters do not always know their motivation. There is not always clear movement.

What I am saying as a feminist, someone who is black would also say: "We have only one form of theater in America." If what you do is not in that form, they say it is not theater. That is why Ntozake Shange called her plays "choreopoems," because they would not let her call them theater. You cannot violate the conventions. I feel there is some stranglehold that is still there. So maybe there is a closer relationship between social science and fiction, or plays, than we think. Maybe they both are last bastions.

I still find that play writing is more the area for me than social science is because the raw material is emotion and character. Social science is about research and finding the truth, and about objectivity. That is why it keeps people from being more personal. I always wanted to be a writer. I just picked academics to make a living. I was hesitant to do social science, at first, because I was afraid it would ruin my writing. I really felt getting a Ph.D. was a cop-out for me. I still think a good novel will tell you more. I believe in creative writing for getting at the truth of how things are. The irony is, that is what social science is supposed to be about.

If social science did not have the conventions, we could do whatever we wanted. I think that would be intriguing. But what would we have then? Some people would probably still want to do the little articles. Other people would be writing poems. You would have to judge each thing on the quality of the work. Would it still be social science if you could do whatever you wanted? Maybe it would not make any difference. I have to think, What are people holding onto when they talk about how it has to be this one way? It's their jobs and their identity. If they could not say, "This is how we do it," would they lose their funding? How would people get tenure and promotion? How would they be judged? I am all for experimenting with the forms, but I see anarchy.

I am faced personally with a dilemma. It is hard to do both. I have been thinking of having to give up doing social science research in order to make a commitment to play writing, at least for some period of time. It bothers me. I think it would be neat if I could put my creativity into my social science. But I really felt that I had to develop in a different field. I wish I could merge the two. I would feel more integrated. I found I just got bored with social science. I got tired of it. However, I feel I would be throwing away twenty years of my life to abandon social science research and writing. I have been working in the area of gender and communication since 1971. I would hate to throw that away, even though I could use it in a play. The social science has been a source of pride and achievement for me. In my play writing, I am trying to learn this new way of writing and it is very difficult. Saying everything in action and dialogue is difficult. But the idea of people speaking your lines: I find that just wonderful.

I think a lot about the split. I don't like the dichotomy. It is frustrating to me. I have no one to talk with about it. I actually have three careers: I teach, I do academic research, and I do creative writing. I try to get them to connect. I think that most of what gets published in social science is trash. I don't know how much closer I can get in form to finding a way that I can feel satisfied with in social science. In my play writing, I am writing this dramatic material based on ethnographic research. I am also trying to study the process of doing it at the same time. But I would rather just do it than talk about how to do it.

I think more women are going to feel dissatisfied with the forms of social science than men. The men are more assured of getting their perqs, so they just go with the system and do not question it as much. I also think the forms of social science just do not resonate with women's experience as much as they do with men's. Women question these forms because they are not how we would normally, or naturally, do things. That is not to say that women do not do it the traditional way. Women learn how to do it the way the game is played, the way the boys do it. But I think there are more women who are going to question the traditional approach, especially feminists. They are going to want to do what comes naturally, and this ain't it.

Ruth: "I am also writing about my own fear of exposure and my fear of disappointing people."

I first learned to write as a journalist. Then in college, in my English classes, I got very strict training in getting rid of adverbs and adjectives, being less descriptive and more critical, and using proper syntax and grammar. In psychology class, I was taken to task because I did not write like a social scientist. I was told to write in the passive intransitive voice and to have no presence. I had thought I was going to be a writer, but I found that everywhere I turned, my writing was wrong.

One of the reasons I became a historian was that I did well on history papers. That was because history combined a concern with the facts of the social sciences and a concern for style. Having the facts be important gave me something to write about that was not original. This helped me because I did not trust my original feelings or voice. I never could write fiction. I had to have a story to tell. History's having a concern for style meant appealing to a reader, and it incorporated the English class lessons I had been taught, including ideas about developing an argument. I am remembering the conflict I felt between a journalistic style and a scholarly style and, within the scholarly style, between the social sciences and the humanities styles. No one tells you about these conflicts when you are in college. They just tell you you are wrong.

In history, I wrote to tell a scholarly story outside myself, although often it was motivated by an implicit personal question. In my training as a graduate student, I got further feedback about how to do scholarly writing. I was taught about structuring paragraphs and building an argument and being true to evidence. In the historical scholarly style I was learning, I told stories that were motivated by a question I had about the world around me, or about myself, but this question was placed on some other subject. For example, if I had a dilemma about whether to be an academic or a political activist, I wrote a biographical story about someone who had had that dilemma several generations before. If I was curious about the social movements around me, I would go back and look to the past to understand how other people fit into the social movements around them.

When I became exposed to feminist ideas in the early 1970s, I felt permission to make the personal part of my question more explicit. My writing then got couched, literally framed, by a contemporary or personal issue. However, the contemporary issue was only a frame, and inside the frame, the body of the work was presented in a regular scholarly way. Although I was writing my dissertation during the time I was becoming exposed to feminist ideas, my dissertation was not something that I felt I could, in any way, place myself in. It had to be the ultimate in scholarly writing. For an article for a feminist journal that I wrote at the time, and that drew on the dissertation, I did allow myself to at least ask contemporary political questions. However, I did not speak in the first person. Soon after I finished my dissertation, I wrote another historical article for a feminist journal in which I used the first person for the first time, not throughout the article, but definitely in the beginning. Later, in the mid-1980s, when I assigned this second article to a graduate class, students asked how I got away with it, with being so politically present in the article. They thought that was not allowed. Other than that second article, most of my academic writing goes back to the mold of using evidence and writing in proper scholarly English. That is the style that gets published in proper scholarly journals.

Another style would be to speak more personally, even more

so than in my second article. I have only had two experiences with that style. Each was with an extended feminist audience, which is different from my historical profession audience. The first instance was an article I wrote about the university that came from personal experience and was in the first person. I think writing that way could only occur after I had tenure, because I was free to do whatever I wanted. I was beyond being criticized. However, I continue to write in the scholarly mode in order to prove that I can do it as well as anybody. The only other time I wrote in an interpretive personal mode was a recent article on a teaching experience and on student responses in a class. That article used student comments on their experiences, but it was written in the first person, by me, with both a political and a personal intent. I wanted to say, "This happened to me. Do with it what you will. Learn from it." This was a less authoritative style than the traditional storytelling that one does in history.

To meet the demands of my position, to get tenure and to maintain respect, I have had to write in the more scholarly way. I enjoy doing the authoritative storytelling kind of writing. I do not just want to chuck it. For a biography that I am working on now, when I think about "intruding" myself, I do not like it. I feel it is disrespectful to my subject's life as the center of what I am doing. I want to keep separate my life and hers. I feel there is a myth we all agree to in writing history. Most historians agree to a form in which there is an authority that says what happened. However, we all know, "it's so-and-so's interpretation." It is our interpretation. I have always believed it is our telling of the past. But I also am part of the myth. I know that in a preface, or in an introduction, I will raise certain personal issues about why I am telling this story of my subject's life and why I need to. But I do not think that a personal style will restructure the way I write the body of the biography. That would be too major a restructuring of my internal order of things. It would be a different kind of project for me to change from a traditional style to one that integrates an authoritative and a personal approach.

I think some people are more comfortable than I am writing in a personal manner. I am thinking of one recent biography of a well-known feminist, in which the biographer had a chapter in the middle discussing her own feelings and responses. I feel,

"That's not what I want to do." If I had models for doing it in a way that appealed to me, I might be more willing. But I have not seen those models. I am looking forward to reading a lot of biographies, and seeing the conventions, and finding out if there are alternatives. I do not think I am a very original writer. I need something outside myself to talk about. I clam up if I have to generate a new form, or an idea, out of nowhere.

The biography with the middle chapter about the author seemed not a very good model for me because I felt it was unbalanced toward the author. In addition, in placement the personal part intruded. This may be very conventional, but I tend to think of something personal in the beginning or the end of a book, but not smack in the middle. I could imagine a model in which the personal material was not just in one chunk, but along the way. But I have not yet gone back to look at such models to see if they appeal to me, or if I have problems with them.

I am wary of examples that seem to be more about the author than about the evidence. One of the things you are taught in graduate training as one of the worst sins you can commit is "presentism." I do not believe that, but I am still very critical of someone who imposes a present on the past, as opposed to sorting things out and distinguishing "them" from "us."

In history, the trick is, in writing about a world you did not live in, or a person you weren't, there is an important stage of getting outside yourself. You need to get outside your historical preconceptions and experience, and into theirs. The purpose of this is to get someone else's worldview, not yours. Then you might put yourself together to say, "How do I relate to this person?" You might take yourself back in. I think that why we react to some uses of the self is because of a fear that the self will overwhelm the other, the subject. I have had one or two conversations with my therapist about this. Obviously, there is always going to be countertransference. The therapist, or the biographer, cannot completely keep herself out. But she has to be open. You have to do both. You have to separate yourself out to get an idea of who the other person is, and you have to see how you are influencing the other. I think about this a great deal now because I am working on a biography. Maybe it would not come up as strongly for me if I was studying something else.

One thing I am doing in my present study, for the first time, is to keep field notes to record my own feelings and my personal experience of the research. This is to help me separate, or intrude, in order to understand what is going on. In keeping the field notes, I have been influenced by anthropologists and sociologists. I have also been talking with my therapist. Things come to my attention when I least expect it. The first time I spoke publicly about this new biography was in a class on women's autobiography. In response to one student's question, "Are you going to interview your subject's daughter?" I said, "No, I can't." Her daughter had died tragically in an automobile accident. A few minutes later, another student asked me how I was going to deal with my own emotional involvement with my subject. She said that a few moments earlier, when I answered the question about my subject's daughter, my whole tone, demeanor, emotions, affect, and body language had changed. I had become extremely upset, or empathetic. I was clearly involved. I realized she was absolutely right. That was what made me start thinking about the issue of the biographer's involvement.

I took the incident of the daughter's automobile accident into therapy. This incident chokes me up each time I come across it. I talked to my subject's grandson about it. He is one of the children of the woman who was killed. He has had many years to deal with his feelings about her death. I still experience pain when I think of the incident. Two things have come up for me. The first is that I lost a friend in a sudden accident a year or two before I read of my subject's daughter's death. I feel I have been reliving that loss. My subject's loss of her daughter is very evocative of my own loss. Then there is my father's death. That layers on a lot of other deaths. I can see how I am reacting to my own life. I do need to separate out. I also need to find out what my subject's reactions were. The death may not have meant the same thing to her that it means to me. I do not want to play out the daughter's death as a turning point in my subject's life if it is not. This is a way that my field notes are useful.

The field notes began while I was doing my first interviews for the biography, because that was a different experience from reading about my subject's life on paper, which is more traditional for a historian. Doing the interviews with people who knew my subject has raised issues of exposure and accountability for me. I feel it is important for me to sort through my feelings about my research experiences so that they do not influence how I present my material. For example, I now know more about some of the people of my study than they know I know. So there is a way I can feel I am being dishonest in not revealing to them what I know when I speak with them. At other times, I think I know information that would profoundly challenge people's beliefs and positions, and I think I have no right to shake that up. There is "right now" and there is "the book." For now, I let it go. I don't say much of a controversial nature, although some things will come out in the book. There are some issues that are serious, for example, the fact that the subject has a deeply romantic relationship with someone, which certain people are aware of and others are not. This will come out in the biography, and some people will be disturbed. This happens all the time in biographies: people fight over revelations. They say, "No, no, no. They were just friends."

In my research, I have asked for the down side of the story. I have not asked for dirt, but I am saying, "This is too rosy a picture. Everyone says this person was a saint." I am beginning to get a picture of someone who hid herself, who did not let people know the self. I do not know what will become of this. It is the kind of thing I am writing about in the field notes. I am discussing limitations of this kind of personality. I am also writing about my own fear of exposure and my fear of disappointing people.

I do think about eventually writing of my relationship with my subject. It is more a question of where and how, than of whether, to do self-revelatory writing. I think that one thing is acknowledging the self and using it to understand; another question is where you put that. I fully expect to write at least an article about my relationship to my subject. I want to write about what drew me to her and about my fieldwork reactions, but that may be very separate from the biography. I am more aware of my involvement in my research than I have ever been in the past because of feminists talking about this.

I think that attention to the involvement of the self varies with different people. One person's psychic survival may have a lot to do with personalizing and creating the self from within.

That person may focus more on the role of the inner self in creating her work. She may have felt overwhelmed by external creators of her self early on, so she fights off the external barriers and tries to burst through with self. My struggle has to do with the creation of the self from externals, with mobilizing external resources. For me, the external world has always been something that I manipulate to get approval. I was taught to have interaction with others, to play them, to anticipate their responses. Women often learn this. But there was a self.

The students I am training now are being trained very conventionally, even by me, because that is how they succeed. It is a very conservative profession. I do not know the equivalents of the Howie Beckers in history. If some people are experimenting with form, the danger is they cease to be historians.

Let me tell you a story of an interchange. At a workshop I went to, a woman showed a video of a performance of a play she had written about a woman she did research on. At the end, there were questions. A prominent woman historian launched into this woman and tore her apart for fictionalizing, because, to make the play work, she did some dramatic structuring that was not exactly the evidence. This senior woman was adamant that you could not do that, especially in women's history. "If you add anything that is not in the record, then you are not a historian," she said. It is such a cardinal sin to make something up: That is lying. That is fabricating evidence. You can speculate if you couch it, but you have to say that you are imagining. If you become unconventional, the big fear is that you will take liberties with evidence. Then are you doing professional history or not?

It is hard for me to imagine alternative forms, although people write plays and historical novels and computer programs. But changing the form does not necessarily change the use of the self. What might I like to see in terms of other models? I would be doing it if I could tell you what it would look like. I'd be curious to see what it would look like in history, but I have trouble imagining other forms, because I, myself, am very conventional. Women's biographies written by women are sometimes self-reflective, but I do not know of self-reflective accounts by women historians within history. Historians do not teach methods classes. We just apprentice. We emulate the master.

Alicia: "I had been trained in ethnoscience, but what I had inside me was that I was constructing it."

When I wrote my dissertation, anthropology was a science. Your job was to figure out the map in the heads of the people you studied. When I came back from Peru to write up my fieldwork, I had been trained in ethnoscience, but what I had inside me was that I was constructing it. It was this exciting, creative feeling.

In the book I am working on now, I argue that there are three discourses about marriage among the people in a Greek village where I did research over a long period of time, and that these set up different understandings. I describe how people talk about marriage, and I set up models to say, "This is what I know." The models are logical models of how the system works. They are like putting puzzle pieces together, and, all of a sudden, things fall into place. Disparate pieces of a culture all make sense. I see writing as getting it so the puzzle works out right. It is being able to arrange the sequence of ideas so you start out here and end up over there. And if you get off on the wrong step, you can't get over there. A lot of the work lies in figuring out how to make the transitions so it makes sense, so you get from here to there.

My book describes shifts in the village over time, but it is not written as a story of then and now. I would rather weave the two under topics. I start each chapter with a few paragraphs on what the changes in the village were, then I step back and go into my models. I argue that different concepts go with the three discourses. A lot of it is in the first person. I say, "So-and-so told me," and, "I think." A lot of the writing came very naturally.

With the organization of this book, I am stating that I am talking about what I have heard people say. I am not claiming that I know what they did or what they believed, which is different from what has traditionally been done in anthropology. Anthropologists were brought up that you wrote, "The Tonga Tonga believed that the gods jump around and dance in the sky." I am saying, "So-and-so told me the gods jump around and dance in the sky."

After the section on the models, which is me, "me and my

model building," I go into more detail about the way life was in the previous period, sticking closely to what people said, and then I go into the changes. One big problem with the book that I have to deal with now is that I think what people say is wrong. I think women have tremendous power in getting married, for instance, but people say they do not. I want to say, "This is how they talk about it, but if you analyze the discourse, what you find is that marriage gives women a whole lot of moral ground from which to have authority. Or you find that men, because they have to go out and show themselves virile and powerful, end up wasting the family money."

People who read my book say, "But that is you saying it is not true. It is you claiming to know what is true in an objective sense." However, that is not what I am saying. I am saying that I also have only what I hear people say. Given that, I am trying to figure out, Why does it make sense to say the things they do? That is why I separate my models in the beginning, to really just be me in that part, and distinguish it from them. I am trying to articulate my own thinking, as opposed to presenting their speech, which comes later on. Where I bought that it is the ethnographer who creates the understanding is in terms of reasoning: it is not my emotions that structures the discussion, it is my reasoning.

The book I wrote before this was on marriage systems among central African tribes. That book felt like a big theoretical breakthrough for me in developing my own way of interpretation. The tribes book was me thinking through how I think about culture and social systems. That book was the biggest thing for me. I used "I." It came naturally because I was criticizing other anthropologists. I could write, "So-and-so said, but it seems to me." What I wrote first was this long manuscript that will never see the light of day, in which I played with the tribes material. That manuscript has my thinking process about patterns and tribal cases. I read everything I could, trying to make sense of it. I wanted to argue, again, that what people said was wrong, that anthropologists misunderstood the system, that they focused on what the tribes people said on the surface. What I did was to sort through why people had it all wrong. I wrote three hundred single-spaced pages on the computer. It started out as a short paper. It is clear to me that my tribes manuscript was me rummaging through, me thinking things out.

There was a big break in my life at the end of the year I wrote that manuscript. Then it took me years to get back. When I finally did rewrite it, I had to take out a lot of the specific tribes material because of the politics of anthropology. I did not want the book to go out simply as a tribes book because of controversies among anthropologists who study the area. Then I realized that what I was doing was building models. That is what I had cared about the material for. I was not writing a historically specific description. What I ended up doing, when I rewrote, was to focus on the second section. I wanted to say, "What are the systems that make this particular form of organization look real?"

In anthropology, I guess I do feel all alone. It is out of synch to be doing models now, even if my models are not the traditional integrated, functionalist kind. In content, what I am saying runs counter to what most people have done in anthropology. It is threatening to people, I think, because if they accept my models, it will mean that they have got it all backward. In the tribes book, for instance, I am arguing that it is not nomadism that causes the pattern. Men do not naturally biologically fight each other over women. Generally, my argument is that the way you understand culture is to understand power, even though power is culturally constructed.

In my work, I look for what makes the language make sense. Even if you find that what people say is wrong and backward, you can also find language where they have it right. But that language is a suppressed discourse. I argue that it is suppressed because it does not do you any good. There is no cultural reason to say it. You get lots of power by saying certain kinds of things. Therefore, that is what gets said. For instance, in my tribes book, men talk about how brave they are and how they can fight other men. That matters because it has tremendous status implications. It makes no sense at all for people to bother talking about how women do all the work. That's obvious. Nobody needs to go around commenting on it. It does not do anybody any good. So you even get women telling women, "You have to get married because who will give you meat?" which is crazy, because, in fact, meat is widely distributed in the culture and women do not have to get married. Women, in fact, do not want to get married.

My arguments are very feminist, but feminism in anthropology is split between the people who think there are sexually egalitarian societies and the people, like me unfortunately, who think that there are not and have not ever been. The egalitarian view is much more popular.

It was exciting for me to write my tribes book. I love using models. I don't really have anybody who loves my models as much as I do. I use "because" a lot when I write, and people howl at that. It sounds very "cause and effect" and very linear. What I want to say is, things hang together. When a colleague and I initially wrote up our material on hunter-gatherer societies, we would pass drafts back and forth. Every time I would get a draft from her, I would turn it into a very linear, cause-effect thing. She would add these wonderful words and create wonderful verbal images. The final result was a hybrid, but it worked. Things fell into place. When I wrote my tribes book without this colleague, the book had much more of me in it. It had my models. I had to get used to people arguing with me.

In anthropology, you always get the people who say, "But my people don't do that," as if that refutes the whole thing. So I say, "So your people don't do it! Models are to think with. They are not meant to reflect reality. If a model is useful, if it helps you see things, then use it. If it doesn't, throw it out the window." I think my style of writing lets me in for a certain type of opposition, because it is this relatively abstract, social sciency thing. But I like it. I think it redoes anthropology. For example, the hunters' and gatherers' behavior has usually been attributed to the fact that the people are nomadic. However, with the model this colleague and I started working on, that I then developed in the tribes book, we have turned all that around. We have said a wife is the basis of a man's status. The men's fighting does not have to do with nomadism.

My biggest fear is that my tribes book will fall into this vast, dark, silent hole and nobody will ever notice it. I worry about this especially since anthropologists are turning away from models. Anthropologists are turning toward postmodernism and creating collages and illustrative juxtapositions. I feel ambivalent about this trend. At its worst, the collage and "the anthropologist as artist" seems to me like the men reasserting themselves against all the threats that have been made against them. It sets up the "great artist" who creates the collage, as opposed to being something more democratic. On the other hand, some really interesting work is being done by people putting themselves into their ethnographies now in a way that they did not do before. They are talking about the ethnography as a product of an interaction between the ethnographer and people. It is ethnographers writing themselves in, and using other forms, like narratives. It does seem to have opened up ethnography a lot.

However, I am ambivalent about it because, if it states no clear argument, you have no way of knowing whether you believe it or not. When it comes down to it, I do think social science is a science. I think there are better and worse interpretations. I really do like to know what an author is saying, or arguing, and to figure out what the evidence is, and determine, do I buy it?

Feminism has affected my writing in the sense of working on things jointly with other women. We then use the "we," meaning "us real people." But it has not caused confessionalism for me. I think that writing from the viewpoint of feminism challenges anthropology. You cannot write in the abstract because you are challenging the assumptions of people who write in the abstract. I recently collaborated with another woman anthropologist on a book on gender and divisions of labor in anthropological studies. She and I wrote, "We, as women, find that these things do not make sense." In my own books, I know that being a woman, I have been more aware that what I got from people I studied was a function of the interaction, and that it was me who put it into place.

When I think about some of the work that gets done in the postmodernist genre and in feminist work of that type, I think of confessionalism and the people who are in the extreme. They say that ethnography results from their interactions with a person, and they write about their moods and feelings. My book does not do that. It has nothing of my moods and feelings. It has lots of my thoughts and my impressions, but it has only the ones that are germane to what I am talking about. I am thinking of one woman anthropologist who said in a conference paper, "Because my aunt had recently died, I had all these feelings that gave me special understanding of death" among the people she was studying. She went on about "my deep understandings" and "my sorrow," describing a lot of her feelings. My feeling was, "Who could care less?" Maybe somebody who cared about her could care, but it was not illumining the culture. It was just portraying her as this truly feelingful person who was more sensitive than you schmucks out there.

Another anthropologist I heard was better. I still think anthropology is to help you understand other people. What this second person did was to talk about how the people he studied said this thing, and it had always puzzled him. He used his feeling about the death of a close friend to open up another culture to him. Whereas this woman's paper did not. All that it added was gooey sentimentality.

I like arguments. When personal feelings contribute to the argument, then I think it's great. It gives you a deeper understanding. But when it is only "me and my experiences among the folks," and I can't see what it adds, I don't like it. Anthropology has gone through this critique that emphasizes that the basis of our knowledge is the ethnographer, and that knowledge is created in interaction. When people bring their personal experience into play to talk about how particular bodies of knowledge get created, then it seems to me very good and very interesting. But some people then say, all there is is the ethnographer. It has a holier-than-thou attitude, and it seems to me aimed at those of us who wrote abstractly, as if we did not recognize that we were constructing knowledge.

I get very resentful of that because, even when I was writing abstractly in my first book, which began as my dissertation, I had a real sense that I was constructing knowledge. I knew that what I wrote was because of me, the particular place I was in, and what I was able to see and hear. At the time, however, one had to write in an ethnoscience form to get published, using the passive voice. The press I went with changed my writing to active, and it got a lot better. I guess I object to when anthropologists act now as if no one knew it before.

I had a recent experience talking about writing with a

235

woman anthropologist who has used a more postmodernist approach, and who has a chapter on "me and my experiences" at the start of one of her books. She talked about writing as sculpture. She said that, for her, writing was like chipping away at a block of marble to reveal this or that. My vision of writing was more like painting. You prepare the canvas underneath, which, to me, is my models. Then you paint the broad colors of the design, which, to me, is the dominant discourse of what people say. Then you proceed to add tones and colors that change the red into purples and blues, which is all the assumptions that underlie the dominant discourse and the discourses that don't get said.

Thus my vision was one of layering colors on top of colors. It was more constructed and organized than hers. Hers was a vision in which you chip away here and reveal this. Then you go around to the other side and you chip away there. Her vision does not hitch together. Mine was like an argument, and arguments are out. I found this woman's book very interesting though, especially as I read more of it. I really liked it. It is about the construction of self in Thailand. I felt that her metaphor was apt. What emerges from her book is not an argument. It is a layered picture. You end up with a rich picture, and you feel that you have learned a tremendous amount, but it's not like my stuff, which is linear.

Karen: "If the theory is true of someone else's life, that's data.
If I feel it is true of my life, I feel guilty."

I don't think I have ever in my life conducted a study and then felt that I communicated through a journal what I found. I often feel I communicate the real essence best when I give talks. I often use videotapes in presentations on research. I will take a clip of an interview and say, "This person can tell you better than I can." That way, some of the qualitative richness comes through.

I do a great deal of interviewing in the research I do on old people and emotions. I start the interviews saying to the person, "Tell me who so-and-so is." Then I elaborate by saying, "What are you really like?" No one can answer that question, but it breaks the ice, and then they start talking. But none of that response ever gets coded. The tapes then get coded for, "Is the person depressed, anxious, having lots of somatic concerns, concerned about a transition they are about to go through?" I train raters to listen to the tapes and to come up with numbers. Things get rated on a scale from 1 to 10. Then the rating goes into a computer. And that is what comes out. That's the person. So it is really reduced. That reduction is what would get published in a psychology journal. I have never been able to feel that I have represented people in the standard publication form.

I feel that I know a great deal because of doing interviews and spending time with people. All my questions come from exposure to a population. In that way, everything that influences me is qualitative. However, when I go to write it up, in some ways, it becomes very, very empty. It is always empty of the richness. I do not think what I write is without value. It is just very limited.

In talking to people, I would really like to know if, in fact, people reduce their social involvement in old age. If they do, I am interested in the rules that seem to guide that. I see my major line of research as looking at people's social and emotional relationships and how that changes over the life span. Almost all studies support the finding that, with age, people reduce social activity. They talk to other people less. They stay home more and have fewer friends. There are basically two theoretical models that have been proposed to explain this: disengagement theory and activity theory. Disengagement theory contends that, with age, people withdraw from their social worlds, and their social worlds withdraw from them, in a symbolic preparation for death. So people become more emotionally withdrawn; they turn inward. They are less interested in others. They become more egocentric with age.

The second model, activity theory, is a reaction to disengagement theory. It claims that the reduction in social activity that we see among older people is simply the consequence of unfortunate circumstances: people die, so you don't have as many friends; your health leads to restrictions of your opportunities for social interaction; and negative stereotypes and views of older people keep others away from you. Activity theorists argue that withdrawing is not a natural part of old age, that it should be combated, and that people who do best in old age are people who replace all of their lost friends and roles, and remain active and highly involved.

When I was in graduate school, disengagement theory was being attacked from all camps. Activity theory was being touted as "the way," and the bottom line of that theory is that older people are really no different than younger people. Psychologically, they are the same. If we could just treat older people a little nicer and give them a few more opportunities—provide a few more opportunities for these unfortunate old folks—then they would be just like us, young people. And we would all feel better.

Both of these theories, I think, represent distorted views of older people. Both represent a view of older people that is rooted squarely within a decrement model of old age. Gerontologists almost cannot ask a question about old people that is not related to decrement. Decrement means some problem, some deficiency, or loss. Disengagement theorists said initially that old people are flattened and emotionally deadened. Activity theorists said, "No, they're not. They are just like us." It is the "just like us" that is the problem for me.

I have been arguing for something that I have termed selectivity theory. I would never have gotten to this theory had I not done hundreds of interviews with old people myself. I feel I ask different questions because I talk to people, instead of giving questionnaires and tabulating the results. In talking to people, it is very clear to me that they are not at all emotionally flattened and uninterested in social relationships—to some people in their lives. However, there are a whole myriad of prospective social partners they are not interested in, and I think these people very likely and very often include the researcher who is knocking at their door saying, "Can I talk to you?" And the old people say, "Nah." And the researchers say, "See. Withdrawal. Symbolic preparation for death."

I think what happens is that people become more selective in their social partners, and I don't think it begins in old age. We just happened to start to look at it in old people, and that is where we saw it. However, we have not compared the finding with data on younger people, on forty- or twenty-year-olds. In some of my research, I have begun to do that. I have looked longitudinally starting with teenagers, and I find that the decline in social interactions begins in adolescence. You see dramatic drops in the rates of interaction, from adolescence on, in the area of acquaintances. But you see increases in contact with spouses, children, and close friends. Even in relationships where you see declines in contacts, you see increases in emotional closeness. So, yes, a quantitative index of rate of interaction is going down. But no, it is not affecting emotional closeness.

It's as if we see reductions in rates, and from that we have these theories about all sorts of other things that have not even been measured. We assume we know something about emotional closeness that we do not know. I think it is bad science, quantitative or qualitative. The assumptions of gerontology are off, and I think they are off because of a reliance on quantitative methods. We have seen these numbers and counts in study after study, and it is very hard not to have that influence us. It certainly influenced me. I have spent years watching people and recording behaviors. That told me something. But talking to people told me a whole lot more.

I think that gerontologists have decided that older people are not doing as well as younger people because older people are saying different kinds of things. For example, in studies of coping, older people say things in interviews that suggest they are just letting life go by, that they are not taking control of it. When asked, "What do you do about a certain problem?" older people tend to say, "Well, I just accept it. When you get older, that's what happens." Younger people, faced with the same problem, or health situation, say, "Well, I would do this, and I would do that. I would change my diet. I would go to a doctor. I would get this surgery done." They name lots and lots of very action-oriented things they would do. So psychologists conclude that older people are coping less well. Because they use less action-oriented language and because they report fewer coping responses. But you can't make that kind of judgment about people's responses.

I think you get better at lots of things with age. It seems to me if you cope very well, you do not have to do as much, because what you are doing works. The number of coping responses is lower among older people. But if you look at coping effectiveness, or how well people do, it's higher, although coping effectiveness is not generally studied. The gerontologists who make these negative judgments about older people and coping are doing it based on their own personal, internal views of how they would do things. I think that is the crux of how we approach gerontology wrong.

We have approached the study of age in much the same way as we have approached the study of gender differences. We have searched for differences, and when we find a difference, it is almost impossible for us not to call it a decrement, or a deficiency. And then we try to see if we can change it. Our whole view of old people is, "Gee, how sad that is, and what can we do to fix it?" We have this whole discipline of gerontology that is filled with these very noble people who want to help old people. Ninety-nine percent of gerontologists want to help old people look like young people, without ever recognizing that there might be something problematic in that, without asking, Why? Why do we want to do that? For me, it is quite similar to gender. It is the NOW version of equal rights: "Let's make women more like men." It does not question whether that is a life that anybody, male or female, should decide is the good life, or better than another life, or another way of living.

I think some of this happens because we do not think about ourselves. In science, as I think we have all been trained, we bend over backward to pretend the researcher is not even in the room. You are writing down what people do, counting off how many times they look or smile. The unspoken oath is, we are all going to pretend that the researcher is invisible and objective, that the researcher is a machine.

I think I have done that. It is a legitimate scientific question to ask about the involvement of oneself in science. But I find it almost impossible to think about. I rarely tell people why I study aging. I have the feeling, "Well, this is not relevant." When I was out of high school, I got married and had a kid and was divorced, all within four years. Soon after that, I got into an automobile accident. I was in a hospital for a long time. The surgeons said, at first, that I would die and then that I would never walk again. I was placed on a ward of old people in the hospital. Lying there all day, I was struck with how differently they were treated than I was treated. I was being rehabilitated and they were not. I also learned about them and their lives.

Simultaneously, I was taking a basic college psychology course by tape. I was reading in my psychology text about what old people were like and, at the same time, I was having conversations and seeing them. But I think what makes me relate to old people, and understand why people might behave in ways that might not look adaptive to an objective observer, is that I saw how it felt. For instance, do you want to sit on the floor and talk to people when you know that if you sit down it takes about an hour to get your knee working again? It takes time to work the joint through. That affects what you might want to do, and who you might want to talk to, and who is worth it. I could see that both models of aging were really off. It was not a disinterest on people's part, it was a cost-benefit assessment. I had one experience, during the early period when I was in the hospital, where I felt I was becoming an old person. I was cared for so much that I lost my ability to think I could do things for myself. I really think there is a personal reason I study old people that dates from that time. But, even now, as I say that, and sometimes people ask me, I think, "Well, I will tell you why I think I study it, but I am not sure if it is true. I can tell you my sense of it, but it is just a reconstruction. There may be nothing to it."

I feel passionately about what I study. Sometimes that embarrasses me, if I am giving a talk, for instance. I guess I feel embarrassed because I feel that people will think I would contaminate what I do, if it has anything to do with me. It should have nothing to do with anything.

When I think about my publications, I think they have to do with my keeping my job, much more than with my worth as a researcher. I don't think I have ever felt good about myself because of a publication. The times when I feel good about what I do related to my research are when I get involved in political action, like with the Gray Panthers, or testify in a court case, or when I do a workshop training on elder abuse. I can draw on the quantitative literature then. I can say, "But that's not true. They are not like that," or, "Research says . . . "

I really do not think the research is worthless, but I always feel there is so much more than the narrow band we report on. I don't think psychologists do much good. I am driven because I think what I study is fascinating. I love talking to people and trying to figure out what they do and how they change. But I feel it is often very selfish. I get all sorts of benefits from my exposure to the people I study. Then I give back this very little bit.

What if people were to want to know more about me and how I arrive at my conclusions? Where do they come from, apart from the methods I use? I feel the selectivity model certainly holds for me. At one point in my life, I was really interested in almost everybody, and now I am really uninterested in lots of people, but very interested in some, much more so than when I was younger. So the model holds for me. But I feel that if I told people that, they would say, "Your theory is biased." If someone else feels that the theory is true of their life, then I see that as data. If I feel it is true of my life, I feel guilty. If it is true of me, that must mean I am biased, or trying to prove that my life is life, for the masses. But that is really laughable. No one theory is going to explain a life.

I think it is amazing to go through life doing what I do, what we do, and not think more about our involvement. How can I not have thought more about how I relate to what I study? My sister is a songwriter, and I think of this line from one of her songs, "With eyes trained as well as yours, you should have seen me sinking fast."

We think we have trained our eyes to really be able to see things. I study emotions and development and adulthood and have not questioned why I do that, or what that does for me, and it's disturbing. I would like to think of myself as more thoughtful than that, and that I would have that information. It's me. Why don't I have that information about me? I could probably give you more information about the subject I study. The training we get in science is to keep yourself out. I would like to think that I would not have totally bought the training, hook, line, and sinker. But I think I have, much more than I know.

Talking about this leaves me with lots of questions: What do I bring to what I am doing, to my research? I feel I need to understand what it is, so I know. My first reaction when I think of writing about myself in relation to the subject I study is, that is not the topic. The topic is not me. It is these people. I feel obliged not to talk about me. If I did, I would write, first off, about why I am interested in old age and who the people are that I am interested in. I would write about how I felt when I first talked to them. However, as soon as I said how I felt, I would feel I had to justify it.

I always think I am going to write a very personal book about the people I have met. I think of Meyerhoff's *Number Our Days,* the form of the book. It is wonderful to read it. I am as interested in her as in her subject. I identify with her. I know those experiences: trying to get people to talk to you and they treat you like a kid. You wonder, "Am I their granddaughter?" and, "Can I get them to be objective?" You say that to them and they say, "C'mon. That's not important." I can relate to that so much that it is fascinating and it's wonderful. I think Meyerhoff's presentation of the people comes so alive. It is so much more vital than these numbers that come out in quantitative analyses that I think it is much better.

I bought the traditional methods when I got trained. I can feel that being part of me. I really do feel my hair standing on end when I start to think, Can I write like that, in a more personal way? I start feeling that I do not have the right to say what I think and feel, because that does not really matter. I want to write a book someday on selectivity theory, where I present my model. Then every other chapter would be about a person I know and that person's life. What about writing about my involvement with them? I wonder, Would anybody be interested? It raises the whole repression issue.

A Challenge

A Concluding Comment

I listen to these interview accounts and I think of O'Keeffe: "I decided that I was a very stupid fool not to at least paint as I wanted to."[2] I think of one of Trimble's Pueblo potters: "A woman can make anything, any kind of shape with her own hands."[3] And I think, "And what of us?"

"I think ethnography takes guts," says Ginger. "But play writing takes a lot of guts."

"To meet the demands of my position, I had to write in the more scholarly way," says Ruth.

"Around me, people, even if they have tenure, suffer if they do not write in this way. They suffer disdain from colleagues" (Kathleen).

"That manuscript was me rummaging through, me thinking things out" (Alicia).

"The people who object to the personal material are not especially articulate about their objections. They just don't like it. They think it is self-indulgent, narcissistic, unnecessary, and superfluous" (Myra).

"I have felt there is something wrong with me" (Diana).

"The unspoken oath is, we are all going to pretend that the researcher is invisible and objective, that the researcher is a machine" (Karen).

"Historians do not tend to use the 'I.' However, great historians (many of whom are men) do use the 'I' " (Eleanor).

"I wanted to say, 'The only reason I did all this was to tell you what I think of it and what it provokes in me' " (Kathleen).

"Those of us who keep breaking the rules—to be creative, to say, 'This is me'—we are also trying to say, 'You know, I am really part of this system too. I want you to accept me, but not at the cost of loss of my integrity, or loss of my sense of self' " (Diana).

I am saying, with these quotes, and with stories of myself, and with tales of a painter and of Pueblo potters, that I think there is a challenge and it does not lie in an abstraction called social science, nor in the nature of academic institutions or a male power structure. The central challenge is closer to home. It lies in what each of us chooses to do when we represent our experiences. Whose rules will we follow? Will we make our own? What is the nature of the self, the "I," that so many of our prohibitions bury? How can we unearth some of the inner worlds that we learn so very well to hide? Are we willing to do this within social science? Do we, in fact, have the guts to say, "You may not like it, but here I am."

Notes

Introduction

1. Robert Nisbet, *Sociology as an Art Form* (New York: Oxford University Press, 1976), the classic work on this subject in sociology, emphasizes the unity of art and science. I have found this volume very useful for providing long-term perspective.

2. Susan Krieger, *Hip Capitalism* (Beverly Hills, Calif.: Sage, 1979); and Krieger, *The Mirror Dance: Identity in a Woman's Community* (Philadelphia: Temple University Press, 1983).

1. Self and Context

1. Similar discomforts are discussed at length in Shulamit Reinharz's autobiographical account, *On Becoming a Social Scientist: From Survey Research and Participant Observation to Experiential Analysis* (New Brunswick, N.J.: Transaction Publications, 1988).

2. Susan Krieger, "Cooptation: A History of a Radio Station" (Ph.D. dissertation, Stanford University, 1976).

3. This is discussed in Susan Krieger, "Research and the Construction of a Text," *Studies in Symbolic Interaction* 2 (1979): 167–187; and in Krieger, "Fiction and Social Science," *Studies in Symbolic Interaction* 5 (1984): 269–286.

3. The Vulnerability of a Writer

1. See Clifford Geertz, *Works and Lives: The Anthropologist as Author* (Stanford: Stanford University Press, 1988); Geertz, "Blurred Genres: The Refiguration of Social Thought," in *Local Knowledge: Further Essays in Interpretive Anthropology* (New York: Basic Books, 1983), pp. 19–35; John Van Maanen, *Tales of the Field: On Writing Ethnography* (Chicago: University of Chicago Press, 1988); James Clifford and George E. Marcus, eds., *Writing Culture: The Politics and Poetics of Ethnography* (Berkeley: University of California Press, 1986); James Clifford, "On Ethnographic Authority," *Representations* 1:2

(1983): 118–146; George E. Marcus and Michael M. J. Fischer, *Anthropology as Cultural Critique: An Experimental Moment in the Human Sciences* (Chicago: University of Chicago Press, 1986); Howard S. Becker, Michal M. McCall, and Lori V. Morris, "Theatres and Communities: Three Scenes," *Social Problems* 36:2 (1989): 93–116; Donald N. McCloskey, *The Rhetoric of Economics* (Madison: University of Wisconsin Press, 1985); John S. Nelson, Allan Megill, and Donald N. McCloskey, eds., *The Rhetoric of the Human Sciences: Language and Argument in Scholarship and Public Affairs* (Madison: University of Wisconsin Press, 1987); Joseph R. Gusfield, "The Literary Art of Science: Comedy and Pathos in Drinking-Driver Research," in *The Culture of Public Problems: Drinking-Driving and the Symbolic Order* (Chicago: University of Chicago Press, 1981), pp. 83–108; Donald P. Spence, *Narrative Truth and Historical Truth: Meaning and Interpretation in Psychoanalysis* (New York: Norton, 1982); Hayden White, *Metahistory: The Historical Imagination in Nineteenth-Century Europe* (Baltimore: Johns Hopkins University Press, 1973); White, *Tropics of Discourse: Essays in Cultural Criticism* (Baltimore: Johns Hopkins University Press, 1978); and Richard Harvey Brown, *A Poetic for Sociology: Toward a Logic of Discovery for the Human Sciences* (New York: Cambridge University Press, 1983).

2. Subjective bases for studies are discussed, in interesting ways, in Patricia A. Adler and Peter Adler, *Membership Roles in Field Research* (Newbury Park, Calif.: Sage, 1987); Renato Rosaldo, *Culture and Truth: The Remaking of Social Analysis* (Boston: Beacon, 1989); Shulamit Reinharz, *On Becoming a Social Scientist: From Survey Research and Participant Observation to Experiential Analysis* (New Brunswick, N.J.: Transaction Publications, 1988); David N. Berg and Kenwyn K. Smith, eds., *The Self in Social Inquiry: Researching Methods* (Newbury Park, Calif.: Sage, 1988), especially Clayton Alderfer, "Taking Our Selves Seriously as Researchers," pp. 35–70, and Alan Peshkin, "Virtuous Subjectivity: In the Participant-Observer's I's," pp. 267–282; Jennifer C. Hunt, *Psychoanalytic Aspects of Fieldwork* (Newbury Park, Calif.: Sage, 1989); Carolyn Ellis, "Sociological Introspection and Emotional Experience," *Studies in Symbolic Interaction* 14:1 (1991); Jean Briggs, "Kapluna Daughter," in Peggy Golde, ed., *Women in the Field: Anthropological Experiences*, 2d ed. (Berkeley: University of California Press, 1986), pp. 19–44; Briggs, "In Search of Emotional Meaning," *Ethos*, 15:1 (1987): 8–15; Dorinne K. Kondo, *Crafting Selves: Power, Gender, and Discourses of Identity in a Japanese Workplace* (Chicago: University of Chicago Press, 1990); Paul Rabinow, *Reflections on Fieldwork in Morocco* (Berkeley: University of California Press, 1977); Arlene Kaplan Daniels, "Self-Deception and Self-Discovery in Fieldwork," *Qualitative Sociology* 6:3 (1983): 195–214; Marianne A. Paget, "Life Mirrors Work Mirrors Text Mirrors Life . . . ," *Social Problems* 37:2 (1990): 137–148; David M. Hayano, "Auto-Ethnography: Paradigms, Problems, and Prospects," *Human Organization* 38:1 (1979): 99–104; R. Ruth Linden, *Making Stories, Making Selves: Writing Sociology after the Holocaust* (Columbus: Ohio State University Press, 1991); and Samuel Baron and Carl Pletsch, eds., *Introspection in Biography: The Biographer's Quest for Self-Awareness* (Hillsdale, N.J.: Analytic Press, 1985).

Several recent feminist collections also contain pertinent discussions of subjectivity, for example, Mary McCanney Gergen, ed., *Feminist Thought and the Structure of Knowledge* (New York: New York University Press, 1988); Sandra Harding, ed., *Feminism and Methodology: Social Science Issues* (Bloomington: Indiana University Press, 1987); Harding, *The Science Question in Feminism* (Ithaca, N.Y.: Cornell University Press, 1986); Gloria Bowles and Renate Duelli Klein, eds., *Theories of Women's Studies* (London: Routledge and Kegan Paul, 1983); Golde, *Women in the Field;* and Evelyn Fox Keller, *Reflections on Gender and Science* (New Haven: Yale University Press, 1985).

3. For exceptions to the tone of remote authority, see especially Rosaldo, *Culture and Truth,* and Barbara Myerhoff, *Number Our Days* (New York: Simon and Schuster, 1980). A telling statement of frustration with social science writing can be found in Mary Louise Pratt, "Fieldwork in Common Places," in Clifford and Marcus, *Writing Culture,* p. 33: "Ethnographic writing tends to be surprisingly boring. How, one asks constantly, could such interesting people doing such interesting things produce such dull books? What did they have to do to themselves?"

4. The Presence of the Self

1. The symbolic interactionist tradition in sociology has dealt more specifically with the self as internal than have other sociological traditions. However, here, too, there is an emphasis on the self as reflecting others, or reflecting society, social roles, social context, or external appearances. In Erving Goffman's terms, there has been an emphasis on the "performances" of the self. See Charles Horton Cooley, *Human Nature and the Social Order* (New Brunswick, N.J.: Transaction, 1983); George H. Mead, *Mind, Self, and Society* (Chicago: University of Chicago Press, 1962); Erving Goffman, *The Presentation of Self in Everyday Life* (New York: Doubleday Anchor, 1959); and Goffman, *Stigma: Notes on the Management of Spoiled Identity* (Englewood Cliffs, N.J.: Prentice-Hall, 1963).

2. A useful review of definitions of the self in the Freudian tradition can be found in Frederick J. Levine and Robert Kraus, "Psychoanalytic Theories of the Self: Contrasting Approaches to the New Narcissism," and Joseph W. T. Redfearn, "Terminology of Ego and Self: From Freud(ians) to Jung-(ians)," in Polly Young-Eisendrath and James A. Hall, eds., *The Book of the Self: Person, Pretext, and Process* (New York: New York University Press, 1987), pp. 306–330 and 384–403. See also discussions of the object relations literature in Nancy Chodorow, *The Reproduction of Mothering: Psychoanalysis and the Sociology of Gender* (Berkeley: University of California Press, 1978). For self psychology, see Heinz Kohut, *The Search for the Self: Selected Writings of Heinz Kohut, 1950–1978* (New York: International Universities Press, 1978).

3. See, for example, Dorinne K. Kondo, *Crafting Selves: Power, Gender, and Discourses of Identity in a Japanese Workplace* (Chicago: University of Chicago Press, 1990).

4. Some theoretical accounts are exceptional, however. In sociology, for example, Denzin attempts not to diminish the experience of a unique and inner self, even while he speaks of social context and uses a generalist language. See Norman K. Denzin, *Interpretive Biography* (Newbury Park, Calif.: Sage, 1989); and Denzin, *On Understanding Emotion* (San Francisco: Jossey-Bass, 1984). Cooley, in *Human Nature,* and Denzin both use the term "self-feeling," which I find more to the point than "self-representation" or "self-image."

5. Charles Taylor discusses key features of the modern Western concept of the self in his very thorough *Sources of the Self: The Making of the Modern Identity* (Cambridge: Harvard University Press, 1989). Especially pertinent is his statement on the taken-for-granted nature of "inwardness" in modern Western culture (pp. 111–114). George E. Marcus and Michael M. J. Fischer discuss cultural variability in ideas of self and emotion, in *Anthropology as Cultural Critique: An Experimental Moment in the Human Sciences* (Chicago: University of Chicago Press, 1986), pp. 45–73. A classic anthropological essay on the issue of the self is Marcel Mauss, "A Category of the Human Spirit," *Psychoanalytic Review* 55:3 (1968 [1938]): 457–481. A good theoretical feminist discussion of internal life and relatedness can be found in Nancy J. Chodorow, "Beyond Drive Theory: Object Relations and the Limits of Radical Individualism," and "Toward a Relational Individualism," in *Feminism and Psychoanalysis* (New Haven: Yale University Press, 1989), pp. 114–162. Earlier classic statements of an object relations nature are W. R. D. Fairbairn, *An Object Relations Theory of the Personality* (New York: Basic Books, 1952); Melanie Klein, *Love, Guilt, and Reparation and Other Works, 1921–1945* (London: Hogarth, 1975); and Edith Jacobson, *The Self and the Object World* (London: Hogarth, 1965). I find the object relations tradition most interesting for understanding the self and self-other relationships.

6. Arlie Hochschild describes female emotion work and status deference in very interesting ways in "Gender, Status, and Feeling," in *The Managed Heart: Commercialization of Human Feeling* (Berkeley: University of California Press, 1983), pp. 162–184. For gender differences in emotion work and in self-other experience as these affect self-experience, see also Carol Gilligan, *In a Different Voice: Psychological Theory and Women's Development* (Cambridge: Harvard University Press, 1982); Nancy Chodorow, "Family Structure and Feminine Personality," in Michelle Zimbalist Rosaldo and Louise Lamphere, eds., *Woman, Culture and Society* (Stanford: Stanford University Press, 1974), pp. 43–66; Chodorow, *Reproduction of Mothering;* Pamela M. Fishman, "Interaction: The Work Women Do," in Barrie Thorne, Cheris Kramarae, and Nancy Henley, eds., *Language, Gender and Society* (Rowley, Mass.: Newbury House, 1983), pp. 89–101; Polly Young-Eisendrath, "The Female Person and How We Talk about Her," in Mary McCanney Gergen, ed., *Feminist Thought and the Structure of Knowledge* (New York: New York University Press, 1988), pp. 152–172; Nancy Mairs, *Plaintext: Deciphering a Woman's Life* (New York: Harper and Row, 1986); and Ursula K. LeGuin, "Bryn Mawr Commencement Ad-

dress," in *Dancing at the Edge of the World: Thoughts on Words, Women, Places* (New York: Grove Press, 1989), pp. 147–160.

7. Jane Flax discusses the painful and disabling nature of a truly fragmented self in *Thinking Fragments: Psychoanalysis, Feminism, and Postmodernism in the Contemporary West* (Berkeley: University of California Press, 1990), pp. 218–219; Flax also provides a good general discussion of the need for ideas of a core self, inner cohesion, or "deep" subjectivity, even in an age of postmodernism, in *Thinking Fragments,* pp. 216–232. See, too, discussions of the self as organizing, and of incoherent and fragmented inner states as terrorizing, in Jane Loevinger, "The Concept of Self or Ego," and Donald P. Spence, "Turning Happenings into Meanings: The Central Role of the Self," in Young-Eisendrath and Hall, *Book of the Self,* pp. 88–94 and 131–150.

8. Kondo, *Crafting Selves,* pp. 3–48, argues for a decentered, fragmented view of the self, as do postmodernist visions generally. See also Gloria Anzaldúa, *Borderlands/La Frontera: The New Mestiza* (San Francisco: Spinsters/Aunt Lute, 1987); and Sandra Harding, *The Science Question in Feminism* (Ithaca, N.Y.: Cornell University Press, 1986), pp. 163–196, on "fractured identities." While, as a structural form, a totally fragmented self may be intolerable or nonfunctional, as a way of describing multiple aspects of a self, ideas of fragmentation can be useful. Fragmentation ideas seem to me clearly useful when they reflect concrete social experiences (e.g., Anzaldúa) and less useful when they are mainly extensions of postmodern rhetoric (as described in Flax).

9. My emphasis in this volume differs from that of contemporary anthropologists concerned with the self in that I seek primarily to indicate how looking at a researcher's self-experience can help to enrich understanding. Anthropologists, in recent works, also have this interest, but they tend to be more often concerned with the self as ethnic or cultural identity, with unmasking authorial selves in texts, and with revealing the selves of researchers as well as of researched in order to counter dominance-subordinance relationships. For recent anthropological trends, see James Clifford and George E. Marcus, eds., *Writing Culture: The Poetics and Politics of Ethnography* (Berkeley: University of California Press, 1986); James Clifford, *The Predicament of Culture: Twentieth-Century Ethnography, Literature, and Art* (Cambridge: Harvard University Press, 1988); Marcus and Fischer, *Anthropology as Cultural Critique;* and Clifford Geertz, *Works and Lives: The Anthropologist as Author* (Stanford: Stanford University Press, 1988). For accounts of how feminists dealing with self and experience differ from male "new ethnographers," see Francis E. Mascia-Lees, Patricia Sharpe, and Colleen Ballerino Cohen, "The Postmodernist Turn in Anthropology: Cautions from a Feminist Perspective," *Signs* 15:1 (1989): 7–33; and Marilyn Strathern, "An Awkward Relationship: The Case of Feminism and Anthropology," *Signs* 12:2 (1987): 276–292. Barbara Christian provides a complementary account for the case of black female experience not comprehended by the new literary critics, in "The Race for Theory," *Feminist Studies* 14:1 (1988): 67–79.

10. Renato Rosaldo discusses the positioned self in *Culture and Truth:*

The Remaking of Social Analysis (Boston: Beacon, 1989), pp. 1–21, as do feminists concerned with "standpoint epistemologies," for example, Nancy Hartsock, "The Feminist Standpoint: Developing the Ground for a Specifically Feminist Historical Materialism," in Sandra Harding and Merrill Hintikka, eds., *Discovering Reality: Feminist Perspectives on Epistemology, Metaphysics, Methodology and Philosophy of Science* (Dordecht, Holland: D. Reidel, 1983), pp. 283–310; and Harding, *Science Question in Feminism*, pp. 136–162. My own view, however, is really less about positions than about internal sensibility.

11. For good examples of the self as a source of fruitful ideas, see Jean Jackson, "On Trying to Be an Amazon," in Tony Larry Whitehead and Mary Ellen Conaway, eds., *Self, Sex, and Gender in Cross-Cultural Fieldwork* (Urbana: University of Illinois Press, 1986), pp. 263–274; Jean Briggs, "Kapluna Daughter," in Peggy Golde, ed., *Women in the Field: Anthropological Experiences*, 2d ed. (Berkeley: University of California Press, 1986), pp. 19–44; Marianne A. Paget, "Life Mirrors Work Mirrors Text Mirrors Life . . . ," *Social Problems* 37:2 (1990): 137–148; Patricia J. Williams, "On Being an Object of Property," *Signs* 14:4 (1988): 5–24; Laura Bohannon [Elenore Smith Bowen], *Return to Laughter* (New York: Doubleday Anchor, 1964); and Karen McCarthy Brown, *Mama Lola: A Voodoo Priestess in Brooklyn* (Berkeley: University of California Press, 1991). Barrie Thorne has put the matter very nicely in speaking of writing of her feelings during fieldwork: "What I have come to see . . . is that there is a deep logic to this way of writing, that these personal experiences were neither confessions, minor preliminaries, nor mere 'how it was done' appendages to the main study, but were closely tied to, and even generative of, the study and its substantive findings" (Thorne, "Political Activist as Participant Observer: Conflicts of Commitment in a Study of the Draft Resistance Movement of the 1960s," in Robert M. Emerson, ed., *Contemporary Field Research: A Collection of Readings* [Boston: Little, Brown, 1983], pp. 231–232).

5. Speaking of Writing

1. Discussing writing has become more public in recent years. See, for example, Howard S. Becker, *Writing for Social Scientists: How to Start and Finish Your Thesis, Book, or Article* (Chicago: University of Chicago Press, 1986); Linda Brodkey, *Academic Writing as Social Practice* (Philadelphia: Temple University Press, 1988); Laurel Richardson, *Writing Strategies: Reaching Diverse Audiences* (Newbury Park, Calif.: Sage, 1990); Richardson, "The Collective Story: Postmodernism and the Writing of Sociology," *Sociological Focus* 21:3 (1988): 199–208; and John Van Maanen, "Trade Secrets: On Writing Ethnography," (Paper presented to the conference on Writing the Social Text, University of Maryland—College Park, November 1989); and Van Maanen, ed., "Special Issue: The Presentation of Ethnographic Research," *Journal of Contemporary Ethnography* 19:1 (1990).

2. Susan Krieger, "Beyond 'Subjectivity': The Use of the Self in Social Science," *Qualitative Sociology* 8:4 (1985): 309–324.

3. Carolyn G. Heilbrun, *Writing a Woman's Life* (New York: Norton, 1988). I speak here of forms, but the term "narrative structure" might also be used. In social science, there has been considerable recent interest in narrative. The discussion I have found most helpful, along with Hayden White, *Metahistory: The Historical Imagination in Nineteenth-Century Europe* (Baltimore: Johns Hopkins University Press, 1973), is Donald P. Spence, *Narrative Truth and Historical Truth: Meaning and Interpretation in Psychoanalysis* (New York: Norton, 1982). Spence distinguishes the narrative tradition's concern with the "coherence" and "continuity" of an account from the historical, or scientific, tradition's concern with "correspondence" of ideas to data. In a similar vein to Spence are Roy Schafer, "Narration in the Psychoanalytic Dialogue," in W. J. T. Mitchell, ed., *On Narrative* (Chicago: University of Chicago Press, 1981), pp. 25–50; and Donald E. Polkinghorne, *Narrative Knowing and the Human Sciences* (Albany: State University of New York Press, 1988). Also helpful are Laurel Richardson, "Narrative and Sociology," *Journal of Contemporary Ethnography* 19:1 (1990): 116–135; and Victor W. Turner and Edward M. Bruner, eds., *The Anthropology of Experience* (Urbana: University of Illinois Press, 1986), pp. 3–30.

4. I saw a version of Heilbrun's book in manuscript form at the time I spoke with this class.

5. The importance of acknowledging authorial responsibility, even in the context of collaboration, is also discussed in Judith Stacey, *Brave New Families: Stories of Domestic Upheaval in Late Twentieth Century America* (New York: Basic Books, 1990), pp. xiii–xiv and 272–278. Stacey says of two of her main research subjects, who, in important ways, collaborate with her: "Both were right to say that this is not their book, but mine" (p. 273). See also discussions of authorial responsibility in Marjorie L. DeVault, "What Counts as Feminist Ethnography?" (Paper presented at Exploring New Frontiers: Qualitative Research Conference, York University, Toronto, May 1990); and Clifford Geertz, *Works and Lives: The Anthropologist as Author* (Stanford: Stanford University Press, 1988).

6. Neither rhetoric emphasizes a subjective individual view. The social science case is obvious, the feminist case less so. It seems to me that although there is an important element within feminist scholarship that advocates individual subjective expression, the emphasis, when the term "subjective" is used, tends to be on the self as representing standpoints (see chapter 4, note 11), or on the self as an individual instance of a common female experience, rather than on the self as unique. In other words, radical individualism is not often a central feminist value. For discussions of feminist values in social science and of particular dilemmas and virtues of a feminist approach, see Judith Stacey, "Can There Be a Feminist Ethnography?" *Women's Studies International Forum* 11:1 (1988): 21–27; Marjorie L. DeVault, "Talking and Listening from Women's Standpoint: Feminist Strategies for Interviewing and Analysis," *Social Problems* 37:1 (1990): 96–116; DeVault, "Women Write Sociology: Rhetorical Strategies," in Albert Hunter, ed., *The Rhetoric of Social Research: Understood and Believed* (New

Brunswick, N.J.: Rutgers University Press, 1990), pp. 97–110; Judith Stacey and Barrie Thorne, "The Missing Feminist Revolution in Sociology," *Social Problems* 32:4 (1985): 301–316; Sandra Harding, ed., *Feminism and Methodology: Social Science Issues* (Bloomington: Indiana University Press, 1987), especially Dorothy Smith, "Women's Perspective as a Radical Critique of Sociology," pp. 84–96, and Harding, "Introduction: Is There a Feminist Method?" pp. 1–14, and "Conclusion: Epistemological Questions," pp. 181–190; Sandra Harding, *The Science Question in Feminism* (Ithaca, N.Y.: Cornell University Press, 1986); Mary M. Gergen, "Toward a Feminist Metatheory and Methodology in the Social Sciences," in *Feminist Thought and the Structure of Knowledge* (New York: New York University Press, 1988), pp. 87–104; and Patricia Hill Collins, "Learning from the Outsider Within: The Sociological Significance of Black Feminist Thought," *Social Problems* 33:6 (1986): S14–S32. New viewpoints of minority group feminists have not yet been as richly developed in the social sciences as elsewhere; thus literary criticism must often be turned to for ideas. Two useful recent collections are Cheryl A. Wall, ed., *Changing Our Own Words: Essays on Criticism, Theory, and Writing by Black Women* (New Brunswick, N.J.: Rutgers University Press, 1989), and Joanne M. Braxton and Andree Nicola McLaughlin, eds., *Wild Women in the Whirlwind: Afra-American Culture and the Contemporary Literary Renaissance* (New Brunswick, N.J.: Rutgers University Press, 1990).

7. Many feminists do not experience feminism as ideology, or as requiring transcendence of the self, as I do. For many, the true, or real, self is felt to be found within feminism; the individual self is felt as liberated rather than constrained.

8. For an interesting gender-sensitive discussion of vision in the life of an artist, see Mary Lowenthal Felstiner, "Taking Her Life/History: The Autobiography of Charlotte Salomon," in Bella Brodzki and Celeste Schenck, eds., *Life/Lines: Theorizing Women's Autobiography* (Ithaca, N.Y.: Cornell University Press, 1988), pp. 320–337. Robert Nisbet discusses vision in general terms in *Sociology as an Art Form* (New York: Oxford University Press, 1976), pp. 42–67.

6. An Anthropologist and a Mystery Writer

1. Clifford Geertz, *Works and Lives: The Anthropologist as Author* (Stanford: Stanford University Press, 1988). Subsequent references to this volume appear in the text.

2. Carolyn G. Heilbrun, *Writing a Woman's Life* (New York: Norton, 1988). Subsequent references to this volume appear in the text.

3. For association of the personal with confessionalism, see Geertz, *Works,* p. 84; for a link with disease, especially the "Diary Disease," see pp. 89–91; for vague mists, see p. 124. Although speaking of the "I" and of approaches to observation in which the self of the anthropologist is used, Geertz repeatedly eschews the personal, substituting the literary. For example, speaking of Malinowski, he says: "It is, again, essential to see that . . .

negotiating the passage from what one has been through 'out there' to what one says 'back here' is not psychological in character. It is literary" (Geertz, *Works,* p. 78).

4. See also Heilbrun, "Parallel Lives," *Women's Review of Books* 6:8 (May 1989): 8–10, for a review of a biography of Ruth Benedict in which Heilbrun cites Geertz.

5. Howard S. Becker, *Writing for Social Scientists: How to Start and Finish Your Thesis, Book, or Article* (Chicago: University of Chicago Press, 1986), p. 37. Also pertinent to many of the themes of this book is Becker, *Art Worlds* (Berkeley: University of California Press, 1982), significant for its attitude of treating art as ordinary work and in terms of patterns of social organization.

6. See chapter 4, note 6, for references dealing with gender differences in self-experience, and especially with a female relational or other orientation.

7. Self, Truth, and Form

1. Georgia O'Keeffe, *Georgia O'Keeffe* (New York: Viking, 1977), opposite plate 24. Subsequent references to this volume appear in the text, identified by the abbreviation *GOK.*

2. Georgia O'Keeffe, *Georgia O'Keeffe: Art and Letters,* comp. Jack Cowart, Juan Hamilton, and Sarah Greenough (New York: New York Graphic Society Books, 1987), p. 171. Subsequent references to this volume appear in the text, identified by the abbreviation *A&L.*

3. Joan Didion, "Georgia O'Keeffe," in *The White Album* (New York: Simon and Schuster, 1979), p. 128.

4. Laurie Lisle, *Portrait of an Artist: A Biography of Georgia O'Keeffe* (New York: Washington Square Press, 1980), p. 326. Subsequent references to this volume appear in the text, identified by the abbreviated title *Portrait.*

5. Didion, "O'Keeffe," p. 127.

6. Diane Middlebrook speaks of "the desire to make our subjects more like ourselves," in Middlebrook, "The Life of a Good Old Girl," *Hudson Review* 41:3 (1988): 585. For other discussions of biographer/subject relationships and dilemmas of likeness and difference, see Samuel Baron and Carl Pletsch, eds., *Introspection in Biography: The Biographer's Quest for Self-Awareness* (Hillsdale, N.J.: Analytic Press, 1985); Carol Ascher, Louise DeSalvo, and Sara Ruddick, eds., *Between Women: Biographers, Novelists, Critics, Teachers and Artists Write about Their Work on Women* (Boston: Beacon, 1984), especially Carol Ascher, "On 'Clearing the Air': My Letter to Simone de Beauvoir," pp. 85–104; Bella Brodzki and Celeste Schenck, eds., *Life/Lines: Theorizing Women's Autobiography* (Ithaca, N.Y.: Cornell University Press, 1988); Estelle C. Jelinek, ed., *Women's Autobiography: Essays in Criticism* (Bloomington: Indiana University Press, 1980); Susan Groag Bell and Marilyn Yalom, eds., *Revealing Lives: Autobiography, Biography, and Gender* (Albany: State University of New York Press, 1990); Norman K. Denzin, *Interpretive Biography* (Newbury Park,

Calif.: Sage, 1989); and Jacques Derrida, *The Ear of the Other: Otobiography, Transference, Translation,* trans. Avital Ronell (New York: Shocken, 1985).

7. All idiosyncratic punctuation and word use appearing in these letters are from the original.

8. Literary views emphasizing reader interpretation are discussed in Jonathan Culler, *On Deconstruction: Theory and Criticism after Structuralism* (Ithaca, N.Y.: Cornell University Press, 1982); Wolfgang Iser, *The Act of Reading: A Theory of Aesthetic Response* (Baltimore: Johns Hopkins University Press, 1978); Stanley Fish, *Is There a Text in This Class?* (Cambridge: Harvard University Press, 1980). And, from a feminist perspective, by Elizabeth A. Flynn and Patrocino P. Schweickart, eds., *Gender and Reading: Essays on Readers, Texts, and Contexts* (Baltimore: Johns Hopkins University Press, 1986); and Nancy K. Miller, "Changing the Subject: Authorship, Writing, and the Reader," in Teresa de Lauretis, ed., *Feminist Studies: Critical Studies* (Bloomington: Indiana University Press, 1986), pp. 102–120. New ethnographic statements emphasizing interpretation by readers can be found in James Clifford and George E. Marcus, eds., *Writing Culture: The Poetics and Politics of Ethnography* (Berkeley: University of California Press, 1986). A critical response to the new ethnographic "death of the author" perspective can be found in Frances E. Mascia-Lees, Patricia Sharpe, and Colleen Ballerino Cohen, "The Postmodernist Turn in Anthropology: Cautions from a Feminist Perspective," *Signs* 15:1 (1989): 7–33. E. D. Hirsch speaks "in defense of the author" in more literary critical terms in *Validity in Interpretation* (New Haven: Yale University Press, 1967), pp. 1–23. Clearly, I am for acknowledging meanings of an author, and for viewing as important what an author thinks she is doing, or intends to do.

9. WNET/Thirteen, *Georgia O'Keeffe,* videotape (Boston: Home Vision, 1977).

10. Georgia O'Keeffe, *Some Memories of Drawings,* ed. Doris Bry (Albuquerque: University of New Mexico Press, 1988), text for drawing 8.

11. Although not speaking exclusively of grasping inner reality, discussions of subjectivity in the recent "interpretive social science" and feminist scholarship literatures are pertinent. See chapter 3, note 2.

12. Even when she takes up pottery, O'Keeffe sees her art as speech: "I hadn't thought much about pottery but now I thought that maybe I could make a pot, too—maybe a beautiful pot—it could become still another language for me. . . . I rolled the clay and coiled it—rolled it and coiled it. I tried to smooth it and I made very bad pots. [Juan] said to me, 'Keep on, keep on—you have to work at it—the clay has a mind of it's own.' He helped me with this and that and I finally have several pots that are not too bad, but I cannot yet make the clay speak—so I must keep on" (O'Keeffe, *GOK,* opposite last photo).

13. Additional sources used for reference but not directly cited in this chapter are Anita Pollitzer, *A Woman on Paper: Georgia O'Keeffe* (New York: Touchstone, 1988); Lisa Mintz Messenger, *Georgia O'Keeffe* (New York: Metro-

politan Museum of Art, 1988); and Roxanna Robinson, *Georgia O'Keeffe: A Life* (New York: Harper and Row, 1989).

8. From O'Keeffe to Pueblo Potters

1. At this time (fall 1988), I had yet to see the O'Keeffe retrospective. I finally saw the exhibit the following spring.

2. Lisle also comments on the invisibility of O'Keeffe's brush strokes. See Laurie Lisle, *Portrait of an Artist: A Biography of Georgia O'Keeffe* (New York: Washington Square Press, 1980), p. viii.

3. Joan Didion, *Democracy* (New York: Pocket Books, 1985), p. 154.

4. Lisle, *Portrait,* pp. 359–360; this wording of O'Keeffe's response is from a National Public Radio broadcast, KALW/San Francisco, April 13, 1989.

5. Ursula K. LeGuin, "It Was a Dark and Stormy Night; or, Why Are We Huddled about the Campfire?" in W. J. T. Mitchell, ed., *On Narrative* (Chicago: University of Chicago Press, 1981), p. 194.

6. Stephen Trimble, *Talking with the Clay: The Art of Pueblo Pottery* (Santa Fe: School of American Research Press, 1987), p. 100.

7. Stephen Trimble, "Talking With Clay," *New Mexico Magazine* 64:8 (August 1986): pp. 45, 47 (quotes 2–6).

8. Trimble, *Talking,* pp. 105, 29, 29.

9. Ibid., p. 22.

9. Pueblo Indian Potters

1. Ruth L. Bunzel, *The Pueblo Potter: A Study of Contemporary Imagination in Primitive Art* (New York: Columbia University Press, 1929), p. 1. Subsequent references to this volume appear in the text.

2. For a discussion of ideas of personal distinctiveness in non-Western cultures, see John Kirkpatrick, "How Personal Differences Can Make a Difference," in Kenneth J. Gergen and Keith E. Davis, eds., *The Social Construction of the Person* (New York: Springer-Verlag, 1985), pp. 227–240. For background sources dealing with the Pueblo pottery tradition other than those directly cited in this chapter, see Larry Frank and Francis H. Harlow, *Historic Pottery of the Pueblo Indians, 1600–1880* (Boston: New York Graphic Society, 1974); Francis H. Harlow, *Modern Pueblo Pottery, 1880–1960* (Flagstaff, Ariz.: Northland Press, 1977); Ward Alan Minge, *Acomá: Pueblo in the Sky* (Albuquerque: University of New Mexico Press, 1976); J. Richard Ambler, with photography by Marc Gaede, *The Anasazi: Prehistoric People of the Four Corners Region* (Flagstaff: Museum of Northern Arizona, 1984); Betty Toulouse, *Pueblo Pottery of the New Mexico Indians* (Santa Fe: Museum of New Mexico Press, 1977); Barbara A. Babcock, "Clay Voices: Invoking, Mocking, Celebrating," in Victor Turner, ed., *Celebration: Studies in Festivity and Ritual* (Washington, D.C.: Smithsonian Institution Press, 1982), pp. 58–76; Barbara A. Babcock, Guy Monthan, and

Doris Monthan, *The Pueblo Storyteller: Development of a Figurative Ceramic Tradition* (Tucson: University of Arizona Press, 1986); and Sally Price, *Primitive Art in Civilized Places* (Chicago: University of Chicago Press, 1989). Judith Schachter Modell, *Ruth Benedict: Patterns of a Life* (Philadelphia: University of Pennsylvania Press, 1983), pp. 171–178, contrasts Ruth Benedict and Ruth Bunzel in their interpretations of the Zuni. For a discussion of industrial expansion and the commercial exploitation of the "spirit of the American Indian," see T. C. McLuhan, *Dream Tracks: The Railroad and the American Indian, 1890–1930* (New York: Harry N. Abrams, 1985). For additional historical background, see Ramón A. Gutierrez, *When Jesus Came, the Corn Mothers Went Away* (Stanford: Stanford University Press, 1991).

3. Alice Marriott, *María: The Potter of San Ildefonso* (Norman: University of Oklahoma Press, 1948). Subsequent references to this volume appear in the text. Marriott uses accents in Maria's and Julian's names in her book. However, for consistency, I have followed the more common and later usage, thus omitting the accents except in Marriott's title.

4. For a discussion of "women's work" as often done while thinking about other things, or while seeming to be engaged in other activity, see Marjorie L. DeVault, "Doing Housework: Feeding and Family Life," in Naomi Gerstel and Harriet Engel Gross, eds., *Families and Work* (Philadelphia: Temple University Press, 1987), pp. 178–191.

5. These quotations are all Hopi because of nearby ruins where shards are found.

6. Susan Peterson, *The Living Tradition of Maria Martinez* (Tokyo: Kodansha, 1978), p. 94.

7. Ibid., p. 100.

8. Susan Peterson, *Lucy M. Lewis: American Indian Potter* (Tokyo: Kodansha, 1984). Subsequent references to this volume appear in the text.

9. Georgia O'Keeffe, *Georgia O'Keeffe: Art and Letters*, comp. Jack Cowart, Juan Hamilton, and Sarah Greenough (New York: New York Graphic Society Books, 1987), p. 202, italics mine.

10. Joan Didion, *Slouching Towards Bethlehem* (New York: Delta, 1968), p. 136.

11. Stephen Trimble, *Talking with the Clay: The Art of Pueblo Pottery* (Santa Fe: School of American Research Press, 1987). Subsequent references to this volume appear in the text.

12. For an extended discussion of tradition, and of "substantive" tradition as different from a "rationalistic, emancipatory outlook," see Edward Shils, *Tradition* (Chicago: University of Chicago Press, 1981).

13. For discussion of conventions of autobiographical writing, see Carolyn G. Heilbrun, *Writing a Woman's Life* (New York: Norton, 1988); Norman K. Denzin, *Interpretive Biography* (Newbury Park, Calif.: Sage, 1989); and Estelle C. Jelinek, ed., *Women's Autobiography: Essays in Criticism* (Bloomington: Indiana University Press, 1980).

14. For a discussion of a design orientation similar to Bunzel's in its

emphasis on individual expression in the context of convention, see Peter Lane, *Ceramic Form: Design and Decoration* (New York: Rizolli, 1988), on contemporary trends in ceramics in Europe and the United States. Relevant to thinking about a design orientation in social science is Robert Nisbet, *Sociology as an Art Form* (New York: Oxford University Press, 1976), especially pp. 31–41 on themes.

15. Evelyn Fox Keller discusses dominance and submission and war metaphors in the hypothesis-testing tradition in *Reflections on Gender and Science* (New Haven: Yale University Press, 1985), pp. 33–42 and 123–126.

10. Psychotherapy and Pottery Making

1. Otto F. Kernberg, *Borderline Conditions and Pathological Narcissism* (New York: Jason Aaronson, 1975); and Margaret S. Mahler, Fred Pine, and Anni Bergman, *The Psychological Birth of the Human Infant: Symbiosis and Individuation* (New York: Basic Books, 1975).

2. D. W. Winnicott, *Playing and Reality* (New York: Basic Books, 1971).

3. The basic writings of Heinz Kohut are *The Analysis of the Self* (New York: International Universities Press, 1971); *The Restoration of the Self* (New York: International Universities Press, 1977); *The Search for the Self: Selected Writings of Heinz Kohut, 1950–1978* (New York: International Universities Press, 1978); *How Does Analysis Cure?* (Chicago: University of Chicago Press, 1984); and *Self Psychology and the Humanities: Reflections on a New Psychoanalytic Approach* (New York: Norton, 1985). See also Joseph Lichtenberg and Samuel Kaplan, eds., *Reflections on Self Psychology* (Hillsdale, N.J.: Analytic Press, 1983).

4. Carol Gilligan, *In a Different Voice: Psychological Theory and Women's Development* (Cambridge: Harvard University Press, 1982). For feminist statements that advocate Kohut's approach precisely because of its affinities with feminist values, see Kay Knox, "Women's Identity: Self Psychology's New Promise," *Women and Therapy* 4:3 (1985): 57–69; and Judith Kegan Gardiner, "Self Psychology as Feminist Theory," *Signs* 12:4 (1987): 761–780.

5. For example, Polly Young-Eisendrath, "The Female Person and How We Talk about Her," in Mary McCanney Gergen, ed., *Feminist Thought and the Structure of Knowledge* (New York: New York University Press, 1988), pp. 152–172; and Nancy J. Chodorow, *Feminism and Psychoanalysis* (New Haven: Yale University Press, 1989), pp. 154–162.

6. Stephen Trimble, *Talking with the Clay: The Art of Pueblo Pottery* (Sante Fe: School of American Research Press, 1987), pp. 13, 101, and 104.

7. Ruth L. Bunzel, *The Pueblo Potter: A Study of Creative Imagination in Primitive Art* (New York: Columbia University Press, 1929), pp. 20, 23.

8. Ibid., p. 20.

9. Ibid., p. 23.

10. Trimble, *Talking,* p. 25 (quotes 1–3) and p. 28 (quotes 4–7).

11. Susan Peterson, *Lucy M. Lewis: American Indian Potter* (Tokyo: Kodansha, 1984), pp. 44–45.

12. Georgia O'Keeffe, *Georgia O'Keeffe: Art and Letters,* comp. Jack Cowart, Juan Hamilton, and Sarah Greenough (New York: New York Graphic Society Books, 1987), p. 189; phrasing from WNET/Thirteen, *Georgia O'Keeffe,* videotape (Boston: Home Vision, 1977).

13. WNET/Thirteen, *Georgia O'Keeffe.*

14. Bunzel, *Pueblo Potter,* p. 61.

15. O'Keeffe, *Art and Letters,* p. 203.

16. Trimble, *Talking,* p. 93.

11. Experiences in Teaching

1. Susan Krieger, "Students and Teachers," in "Midwestern Stories" (Manuscript, 1978), p. 1.

2. Ibid., pp. 1–2.

3. Ibid., p. 2.

4. Ibid., p. 3.

5. Nancy Mairs, "On Living Behind Bars," and "On Being Raised by a Daughter," in *Plaintext: Deciphering a Woman's Life* (New York: Harper and Row, 1986), pp. 125–154, and pp. 63–76.

6. Student paper 14, Women and Organizations, spring 1989. Subsequent papers cited in this chapter are from this class.

7. Paper 16.

8. Paper 22.

9. Ibid.

10. Paper 15.

11. Paper 2. "Experiences as truth" refers to Ursula K. LeGuin, quoted at start of chapter 12.

12. Paper 15.

13. Paper 6.

14. Paper 22. The "mother tongue" refers to LeGuin: "The mother tongue . . . is primitive: inaccurate, coarse, limited, trivial, banal. It's repetitive, the same over and over, like the work called women's work; earthbound, housebound. It's vulgar, the vulgar tongue, common, common speech, colloquial. . . . The mother tongue is language not as mere communication, but as relation, relationship. It connects. It goes two ways, many ways, an exchange, a network. Its power is not in dividing but in binding, not in distancing but in uniting. It is written, but not by scribes and secretaries for posterity; it flies from the mouth on the breath that is our life and is gone, like the outbreath, utterly gone and yet returning. . . . It is a language always on the verge of silence and often on the verge of song. It is the language stories are told in. It is the language spoken by all children and most women, and so I call it the mother tongue, for we learn it from our mothers and speak it to our kids. I'm trying to use it here in public where it

isn't appropriate, not suited to the occasion, but I want to speak it to you because we are women and I can't say to you what I want to say about women in the language of capital M Man" (LeGuin, "Bryn Mawr Commencement Address," in *Dancing at the Edge of the World: Thoughts on Words, Women, Places* [New York: Grove Press, 1989], pp. 149–150).

15. Susan Krieger, "Fiction and Social Science: A Methodological Inquiry" (Manuscript, 1986), discussed in chapter 3, this volume.

16. Paper 10.

17. See chapters 14 and 15 for other views.

18. I am reminded of Goffman's statement at the end of *Asylums:* "Our status is backed by the solid buildings of the world, while our sense of personal identity often resides in the cracks" (Erving Goffman, *Asylums: Essays on the Social Situation of Mental Patients and Other Inmates* [Garden City, N.Y.: Anchor, 1961], p. 320).

12. Snapshots of Research

1. The question of why people trust an interviewer is discussed in Susan Krieger, "Research and the Construction of a Text," *Studies in Symbolic Interaction* 2 (1979): 167–187.

2. Student statement, seminar on research methods, spring 1983.

3. Joan Didion, *Slouching Towards Bethlehem* (New York: Delta, 1968), p. xiv.

4. In my sense of this, I often think of Virginia Woolf's essay "Mr. Bennett and Mrs. Brown," in which Woolf describes an elderly lady she catches sight of on a train and her wish to capture this woman's character. Yet her Mrs. Brown escapes her; she is a figure just beyond reach, beckoning with the phrase, "Catch me if you can." Says Woolf of Brown (a character, a reality): "Few catch the phantom; most have to be content with a scrap of her dress or a wisp of her hair" (Virginia Woolf, "Mr. Bennett and Mrs. Brown," in *The Captain's Death Bed and Other Essays* [New York: Harcourt Brace Jovanovich, 1978], p. 94). See also Judith Stacey's "Epilogue: Taking Women at Their Word," in *Brave New Families: Stories of Domestic Upheaval in Late Twentieth Century America* (New York: Basic Books, 1990), pp. 272–278, for a discussion of a research subject's response to attempts to capture her in a study.

5. Susan Krieger, *The Mirror Dance: Identity in a Women's Community* (Philadelphia: Temple University Press, 1983), p. 154.

6. At the time, I thought the book would be called "House for Sale: Professionalism among Women Real Estate Agents."

7. Susan Krieger, "Jenny's World" (Manuscript, 1985).

8. Susan Krieger, "Fiction and Social Science: A Methodological Inquiry" (Manuscript, 1986).

9. Descriptions of interviewing as experienced by others may be useful for perspective. I have found particularly interesting statements on interviewing in Marjorie L. DeVault, "Talking and Listening from Women's Standpoint:

Feminist Strategies for Interviewing and Analysis," *Social Problems* 37:1 (1990): 96–116; Marianne A. Paget, "Experience and Knowledge," *Human Studies* 6 (1983): 67–90; Ann Oakley, "Interviewing Women: A Contradiction in Terms," in Helen Roberts, ed., *Doing Feminist Research* (London: Routledge and Kegan Paul, 1981), pp. 30–61; Harry Stack Sullivan, *The Psychiatric Interview* (New York: Norton, 1970); and Everett C. Hughes, "Of Sociology and the Interview," in *The Sociological Eye: Selected Papers* (New Brunswick, N.J.: Transaction Books, 1984), pp. 507–515. Says Hughes of the interview, "But the interview is still more than a tool. . . . It is our flirtation with life, our eternal affair" (Hughes, *Sociological Eye,* p. 508).

13. Beyond Subjectivity

This article originally appeared in *Qualitative Sociology* 8:4 (1985): 309–324; it is used here by permission. For their help in preparing the original article, I thank Estelle Freedman, Marythelma Brainard, Nancy Chodorow, Meredith Gould, and Ann Swidler.

1. This restimulation of interest has been sparked most dramatically by the development of feminist scholarship across fields. This new scholarship has led to a reexamination not only of the difference that gender makes in determining what we see, and how we see it, but of other perceptual nets as well. In the recent literature, of great interest are Evelyn Fox Keller's writings on gender and science: "Feminism and Science," *Signs* 7:3 (1982): 589–602; "Feminism as an Analytic Tool for the Study of Science," *Academe* 69:5 (1983): 15–21; *A Feeling for the Organism: The Life and Work of Barbara McClintock* (San Francisco: W. H. Freeman, 1983); and *Reflections on Gender and Science* (New Haven: Yale University Press, 1985). Keller deals with notions of objectivity, subject-object splits, and gender in the work of scientists. See also Carol Gilligan, *In a Different Voice: Psychological Theory and Women's Development* (Cambridge: Harvard University Press, 1982), concerning women's distinctive developmental experiences and how these can lead to highly contextual ways of seeing; and Nancy Chodorow, *The Reproduction of Mothering: Psychoanalysis and the Sociology of Gender* (Berkeley: University of California Press, 1978), which provides a basic psychoanalytic statement concerning women's self-other relationships. Each of these works, to a significant degree, draws on theories of object relations, a field in which an important recent contribution is Margaret S. Mahler, Fred Pine, and Anni Bergman, *The Psychological Birth of the Human Infant: Symbiosis and Individuation* (New York: Basic Books, 1975).

For the past ten years, feminist anthropologists have been particularly articulate in encouraging the recognition of gender-related observer biases; a recent overview can be found in Jane Monnig Atkinson, "Review Essay: Anthropology," *Signs* 8:2 (1982): 232–258. Many of the earlier classic questions are framed in Michelle Zimbalist Rosaldo, "The Use and Abuse of Anthropology: Reflections on Feminism and Cross-Cultural Understanding," *Signs* 5:3

(1980): 389–417; Rayna R. Reiter, ed., *Toward an Anthropology of Women* (New York: Monthly Review Press, 1975); and Michelle Zimbalist Rosaldo and Louise Lamphere, eds., *Woman, Culture, and Society* (Stanford: Stanford University Press, 1974). Nonfeminist anthropologists interested in the "new ethnography" have also been concerned specifically with the observer-observed relationship, although in a different vein; see, for example, James Clifford, "Fieldwork, Reciprocity, and the Making of Ethnographic Texts: The Example of Maurice Leenhardt," *Man* 15 (1980): 518–532; Clifford, "On Ethnographic Authority," *Representations* 1:2 (1983): 118–146; Paul Rabinow, *Reflections on Fieldwork in Morocco* (Berkeley: University of California Press, 1977); Rabinow, " 'Facts Are a Word of God': An Essay Review of James Clifford's *Person and Myth: Maurice Leenhardt in the Melanesian World*," in George W. Stocking, Jr., ed., *Observers Observed: Essays on Ethnographic Fieldwork* (Madison: University of Wisconsin Press, 1983), pp. 196–207; and the works of Renato Rosaldo and Clifford Geertz. An astute comparison of differences between the new feminist and nonfeminist anthropologists in their treatment of subject-object can be found in Marilyn Strathern, "Dislodging a World View: Challenge and Counter-Challenge in the Relationship between Feminism and Anthropology" (Draft of a lecture given in the series "Changing Paradigms: The Impact of Feminist Theory upon the World of Scholarship," at the Research Centre for Women's Studies, Adelaide, Australia, 1984). Finally, there is a category of other prominent recent works that either embody or call attention to subject-object and relational issues in a new way, for example, Barbara Myerhoff, *Number Our Days* (New York: Simon and Schuster, 1980); Arlie Russell Hochschild, *The Managed Heart: Commercialization of Human Feeling* (Berkeley: University of California Press, 1983); and Gloria Bowles and Renate Duelli Klein, eds., *Theories of Women's Studies* (London: Routledge and Kegan Paul, 1983).

2. Important discussions specific to sociology can be found in Dorothy E. Smith, "Women's Perspective as a Radical Critique of Sociology," *Sociological Inquiry* 44 (1974): 7–13; Smith, "A Sociology for Women," in Julia A. Sherman and Evelyn Torton Beck, eds., *The Prism of Sex: Essays in the Sociology of Knowledge* (Madison: University of Wisconsin Press, 1979), pp. 135–187; Marcia Millman and Rosabeth Moss Kanter, eds., *Another Voice: Feminist Perspectives on Social Life and Social Science* (New York: Anchor, 1975); Shulamit Reinharz, *On Becoming a Social Scientist: From Survey Research and Participant Observation to Experiential Analysis* (San Francisco: Jossey-Bass, 1979; reprint ed., New Brunswick, N.J.: Transaction Publications, 1988); Meredith Gould, "Review Essay: The New Sociology," *Signs* 5:3 (1980): 459–467; Helen Roberts, ed., *Doing Feminist Research* (London: Routledge and Kegan Paul, 1981); and Judith Stacey and Barrie Thorne, "The Missing Feminist Revolution in Sociology" (Paper presented at the Annual Meetings of the American Sociological Association, San Antonio, 1984; published in *Social Problems* 32:4 [1985]: 301–316).

3. Instructions to field researchers to acknowledge and deal with

contextual effects and with personal roles and biases have long been common in sociological texts on qualitative method. Further, many of our classics in sociology have distinctly personal tones and styles. However, I believe there is something new being said today, and it is being said most prominently by feminist scholars. This new statement concerns both what a personal style can be and what we mean by the term "participant-observation." The feminists, in effect, are trying to point out that, traditionally, we have allowed the "personal" only if it was male; we do not even yet fully know what the female social scientist's voice might be. Further, it has never been at all clear exactly what we mean by participant-observation, but certainly the rational balancing of "distance" and "involvement" that is usually implied is something qualitatively different from what Keller, for instance, means when she speaks of "a feeling for the organism," and, indeed, of "love" (Keller, *A Feeling* and "Feminism as Analytic Tool").

4. Susan Krieger, *The Mirror Dance: Identity in a Women's Community* (Philadelphia: Temple University Press, 1983).

5. I read Gearing on studying Fox Indians: "When one is estranged, he is unable to relate because he cannot see enough to relate to. . . . The opposite of being estranged is to find a people believable" (Fred Gearing, *The Face of the Fox* [Chicago: Aldine, 1970], p. 5).

6. Susan Krieger, "The Group as Significant Other: Strategies for Definition of the Self" (Paper presented at the Annual Meetings of the Pacific Sociological Association, San Francisco, April 1980).

14. Problems of Self and Form, I

1. By "anonymity," I mean only that I have omitted formal identifications (by changing names of persons, institutions, and publications). Individuals, of course, always can be informally identified by some people, but I think the use of an official pseudonym means that identification happens less often than might otherwise be the case. The use of a pseudonym also suggests that the speaker is a fiction—a creation of the researcher and the person interviewed. This created person is different from some other "real person" whom one might seek to view as the source of an account. Finally, I should note that none of my interviewees officially reviewed the interview tales I finally wrote, although I checked a few sensitive matters with several of them. It seems to me that the current climate surrounding innovative ethnographic research favors review by people studied. I do not think such review is wrong, but, for myself, I find it constraining. I feel it is opposite to the idea of what doing a study is for me, which is to try to make my own sense, or portrait, with the material I have gathered, a sense that may very well not be shared by others.

2. This first "Other Voices" chapter is based on interviews conducted during summer 1989 with a sociologist, a historian, an educational sociologist, and a clinical psychologist, presented in that order.

15. Problems of Self and Form, II

1. This chapter is based on interviews conducted during summer 1989 with a communication researcher, a historian, an anthropologist, and a psychologist, presented in that order.

2. Georgia O'Keeffe, *Georgia O'Keeffe* (New York: Viking, 1977), opposite plate 12.

3. Stephen Trimble, *Talking with the Clay: The Art of Pueblo Pottery* (Santa Fe: School of American Research Press, 1987), p. 13.

Bibliography

Adler, Patricia A., and Peter Adler. *Membership Roles in Field Research*. Newbury Park, Calif.: Sage, 1987.

Ambler, J. Richard. Photography by Marc Gaede. *The Anasazi: Prehistoric People of the Four Corners Region*. Flagstaff: Museum of Northern Arizona, 1984.

Anzaldúa, Gloria. *Borderlands/La Frontera: The New Mestiza*. San Francisco: Spinsters/Aunt Lute, 1987.

Ascher, Carol, Louise DeSalvo, and Sara Ruddick, eds. *Between Women: Biographers, Novelists, Critics, Teachers and Artists Write about Their Work on Women*. Boston: Beacon, 1984.

Atkinson, Jane Monnig. "Review Essay: Anthropology." *Signs* 8:2 (1982): 232–258.

Babcock, Barbara A. "Clay Voices: Invoking, Mocking, Celebrating." In *Celebration: Studies in Festivity and Ritual*, ed. Victor Turner, pp. 58–76. Washington, D.C.: Smithsonian Institution Press, 1982.

Babcock, Barbara A., Guy Monthan, and Doris Monthan. *The Pueblo Storyteller: Development of a Figurative Ceramic Tradition*. Tucson: University of Arizona Press, 1986.

Baron, Samuel, and Carl Pletsch, eds. *Introspection in Biography: The Biographer's Quest for Self-Awareness*. Hillsdale, N.J.: Analytic Press, 1985.

Becker, Howard S. *Art Worlds*. Berkeley: University of California Press, 1982.
———. *Writing for Social Scientists: How to Start and Finish Your Thesis, Book, or Article*. Chicago: University of Chicago Press, 1986.

Becker, Howard S., Michal M. McCall, and Lori V. Morris. "Theatres and Communities: Three Scenes." *Social Problems* 36:2 (1989): 93–116.

Bell, Susan Groag, and Marilyn Yalom, eds. *Revealing Lives: Autobiography, Biography, and Gender*. Albany: State University of New York Press, 1990.

Berg, David N., and Kenwyn K. Smith, eds. *The Self in Social Inquiry: Researching Methods*. Newbury Park, Calif.: Sage, 1988.

Bohannon, Laura [Elenore Smith Bowen]. *Return to Laughter*. New York: Doubleday Anchor, 1964.

Bowles, Gloria, and Renate Duelli Klein, eds. *Theories of Women's Studies.* London: Routledge and Kegan Paul, 1983.

Braxton, Joanne M., and Andree Nicola McLaughlin, eds. *Wild Women in the Whirlwind: Afra-American Culture and the Contemporary Literary Renaissance.* New Brunswick, N.J.: Rutgers University Press, 1990.

Briggs, Jean. "In Search of Emotional Meaning." *Ethos* 15:1 (1987): 8–15.

———. "Kapluna Daughter." In *Women in the Field: Anthropological Experiences,* 2d ed., ed. Peggy Golde, pp. 19–44. Berkeley: University of California Press, 1986.

Brodkey, Linda. *Academic Writing as Social Practice.* Philadelphia: Temple University Press, 1988.

Brodzki, Bella, and Celeste Schenck, eds. *Life/Lines: Theorizing Women's Autobiography.* Ithaca, N.Y.: Cornell University Press, 1988.

Brown, Karen McCarthy. *Mama Lola: A Voodoo Priestess in Brooklyn.* Berkeley: University of California Press, 1991.

Brown, Richard Harvey. *A Poetic for Sociology: Toward a Logic of Discovery for the Human Sciences.* New York: Cambridge University Press, 1983.

Bunzel, Ruth L. *The Pueblo Potter: A Study of Contemporary Imagination in Primitive Art.* New York: Columbia University Press, 1929.

Chodorow, Nancy. "Family Structure and Feminine Personality." In *Woman, Culture, and Society,* ed. Michelle Zimbalist Rosaldo and Louise Lamphere, pp. 43–66. Stanford: Stanford University Press, 1974.

———. *Feminism and Psychoanalysis.* New Haven: Yale University Press, 1989.

———. *The Reproduction of Mothering: Psychoanalysis and the Sociology of Gender.* Berkeley: University of California Press, 1978.

Christian, Barbara. "The Race for Theory." *Feminist Studies* 14:1 (1988): 67–79.

Clifford, James. "Fieldwork, Reciprocity, and the Making of Ethnographic Texts: The Example of Maurice Leenhardt." *Man* 15 (1980): 518–532.

———. "On Ethnographic Authority." *Representations* 1:2 (1983): 118–146.

———. *The Predicament of Culture: Twentieth-Century Ethnography, Literature, and Art.* Cambridge: Harvard University Press, 1988.

Clifford, James, and George E. Marcus, eds. *Writing Culture: The Poetics and Politics of Ethnography.* Berkeley: University of California Press, 1986.

Collins, Patricia Hill. "Learning from the Outsider Within: The Sociological Significance of Black Feminist Thought." *Social Problems* 33:6 (1986): S14–S32.

Cooley, Charles Horton. *Human Nature and the Social Order.* New Brunswick, N.J.: Transaction, 1983.

Culler, Jonathan. *On Deconstruction: Theory and Criticism after Structuralism.* Ithaca, N.Y.: Cornell University Press, 1982.

Daniels, Arlene Kaplan. "Self-Deception and Self-Discovery in Fieldwork." *Qualitative Sociology* 6:3 (1983): 195–214.

Denzin, Norman K. *Interpretive Biography.* Newbury Park, Calif.: Sage, 1989.

————. *On Understanding Emotion.* San Francisco: Jossey-Bass, 1984.

Derrida, Jacques. *The Ear of the Other: Otobiography, Transference, Translation.* Trans. Avital Ronell. New York: Shocken, 1985.

DeVault, Marjorie L. "Doing Housework: Feeding and Family Life." In *Families and Work,* ed. Naomi Gerstel and Harriet Engel Gross, pp. 178–191. Philadelphia: Temple University Press, 1987.

————. "Talking and Listening from Women's Standpoint: Feminist Strategies for Interviewing and Analysis." *Social Problems* 37:1 (1990): 96–116.

————. "What Counts as Feminist Ethnography?" Paper presented at Exploring New Frontiers: Qualitative Research Conference, York University, Toronto, May 1990.

————. "Women Write Sociology: Rhetorical Strategies." In *The Rhetoric of Social Research: Understood and Believed,* ed. Albert Hunter, pp. 97–110. New Brunswick, N.J.: Rutgers University Press, 1990.

Didion, Joan. *Democracy.* New York: Pocket Books, 1985.

————. *Slouching Towards Bethlehem.* New York: Delta, 1968.

————. *The White Album.* New York: Simon and Schuster, 1979.

Ellis, Carolyn. "Sociological Introspection and Emotional Experience." *Studies in Symbolic Interaction* 14:1 (1991).

Fairbairn, W.R.D. *An Object Relations Theory of the Personality.* New York: Basic Books, 1952.

Felstiner, Mary Lowenthal. "Taking Her Life/History: The Autobiography of Charlotte Salomon." In *Life/Lines: Theorizing Women's Autobiography,* ed. Bella Brodzki and Celeste Schenck, pp. 320–337. Ithaca, N.Y.: Cornell University Press, 1988.

Fish, Stanley. *Is There a Text in This Class?* Cambridge: Harvard University Press, 1980.

Fishman, Pamela M. "Interaction: The Work Women Do." In *Language, Gender and Society,* ed. Barrie Thorne, Cheris Kramarae, and Nancy Henley, pp. 89–101. Rowley, Mass.: Newbury House, 1983.

Flax, Jane. *Thinking Fragments: Psychoanalysis, Feminism and Postmodernism in the Contemporary West.* Berkeley: University of California Press, 1990.

Flynn, Elizabeth A., and Patrocino P. Schweickart, eds. *Gender and Reading: Essays on Readers, Texts, and Contexts.* Baltimore: Johns Hopkins University Press, 1986.

Frank, Larry, and Francis H. Harlow. *Historic Pottery of the Pueblo Indians, 1600–1880.* Boston: New York Graphic Society, 1974.

Gardiner, Judith Kegan. "Self Psychology as Feminist Theory." *Signs* 12:4 (1987): 761–780.

Gearing, Fred. *The Face of the Fox.* Chicago: Aldine, 1970.

Geertz, Clifford. *Local Knowledge: Further Essays in Interpretive Anthropology.* New York: Basic Books, 1983.

————. *Works and Lives: The Anthropologist as Author.* Stanford: Stanford University Press, 1988.

Gergen, Mary McCanney, ed. *Feminist Thought and the Structure of Knowledge.* New York: New York University Press, 1988.

Gilligan, Carol. *In a Different Voice: Psychological Theory and Women's Development.* Cambridge: Harvard University Press, 1982.

Goffman, Erving. *Asylums: Essays on the Social Situation of Mental Patients and Other Inmates.* Garden City, N.Y.: Anchor, 1961.

————. *The Presentation of Self in Everyday Life.* New York: Doubleday Anchor, 1959.

————. *Stigma: Notes on the Management of Spoiled Identity.* Englewood Cliffs, N.J.: Prentice-Hall, 1963.

Golde, Peggy. *Women in the Field: Anthropological Experiences.* 2d ed. Berkeley: University of California Press, 1977.

Gould, Meredith. "Review Essay: The New Sociology." *Signs* 5:3 (1980): 459–467.

Gusfield, Joseph R. *The Culture of Public Problems: Drinking-Driving and the Symbolic Order.* Chicago: University of Chicago Press, 1981.

Gutierrez, Ramón A. *When Jesus Came, the Corn Mothers Went Away.* Stanford: Stanford University Press, 1991.

Harding, Sandra, ed. *Feminism and Methodology: Social Science Issues.* Bloomington: Indiana University Press, 1987.

————. *The Science Question in Feminism.* Ithaca, N.Y.: Cornell University Press, 1986.

Harlow, Francis H. *Modern Pueblo Pottery, 1880–1960.* Flagstaff, Ariz.: Northland Press, 1977.

Hartsock, Nancy. "The Feminist Standpoint: Developing the Ground for a Specifically Feminist Historical Materialism." In *Discovering Reality: Feminist Perspectives on Epistemology, Metaphysics, Methodology and Philosophy of Science,* ed. Sandra Harding and Merrill Hintikka, pp. 283–310. Dordecht, Holland: D. Reidel, 1983.

Hayano, David M. "Auto-Ethnography: Paradigms, Problems, and Prospects." *Human Organization* 38:1 (1979): 99–104.

Heilbrun, Carolyn G. "Parallel Lives." *Women's Review of Books* 6:8 (May 1989): 8–10.

————. *Writing a Woman's Life.* New York: Norton, 1988.

Hirsch, E. D. *Validity in Interpretation.* New Haven: Yale University Press, 1967.

Hochschild, Arlie Russell. *The Managed Heart: Commercialization of Human Feeling.* Berkeley: University of California Press, 1983.

Hughes, Everett C. *The Sociological Eye: Selected Papers.* New Brunswick, N.J.: Transaction Books, 1984.

Hunt, Jennifer C. *Psychoanalytic Aspects of Fieldwork.* Newbury Park, Calif.: Sage, 1989.

Iser, Wolfgang. *The Act of Reading: A Theory of Aesthetic Response.* Baltimore: Johns Hopkins University Press, 1978.

Jackson, Jean. "On Trying to Be an Amazon." In *Self, Sex, and Gender in*

Bibliography

Cross-Cultural Fieldwork, ed. Tony Larry Whitehead and Mary Ellen Conaway, pp. 263–274. Urbana: University of Illinois Press, 1986.

Jacobson, Edith. *The Self and the Object World*. London: Hogarth, 1965.

Jelinek, Estelle C., ed. *Women's Autobiography: Essays in Criticism*. Bloomington: Indiana University Press, 1980.

Keller, Evelyn Fox. *A Feeling for the Organism: The Life and Work of Barbara McClintock*. San Francisco: W. H. Freeman, 1983.

———. "Feminism and Science." *Signs* 7:3 (1982): 589–602.

———. "Feminism as an Analytic Tool for the Study of Science." *Academe* 69:5 (1983): 15–21.

———. *Reflections on Gender and Science*. New Haven: Yale University Press, 1985.

Kernberg, Otto F. *Borderline Conditions and Pathological Narcissism*. New York: Jason Aaronson, 1975.

Kirkpatrick, John. "How Personal Differences Can Make a Difference." In *The Social Construction of the Person*, ed. Kenneth J. Gergen and Keith E. Davis, pp. 227–240. New York: Springer-Verlag, 1985.

Klein, Melanie. *Love, Guilt, and Reparation and Other Works, 1921–1945*. London: Hogarth, 1975.

Knox, Kay. "Women's Identity: Self Psychology's New Promise." *Women and Therapy* 4:3 (1985): 57–69.

Kohut, Heinz. *The Analysis of the Self*. New York: International Universities Press, 1971.

———. *How Does Analysis Cure?* Chicago: University of Chicago Press, 1984.

———. *The Restoration of the Self*. New York: International Universities Press, 1977.

———. *The Search for the Self: Selected Writings of Heinz Kohut, 1950–1978*. New York: International Universities Press, 1978.

———. *Self Psychology and the Humanities: Reflections on a New Psychoanalytic Approach*. New York: Norton, 1985.

Kondo, Dorinne K. *Crafting Selves: Power, Gender, and Discourses of Identity in a Japanese Workplace*. Chicago: University of Chicago Press, 1990.

Krieger, Susan. "Beyond 'Subjectivity': The Use of the Self in Social Science." *Qualitative Sociology* 8:4 (1985): 309–324.

———. "Cooptation: A History of a Radio Station." Ph.D. dissertation, Stanford University, 1976.

———. "Fiction and Social Science." *Studies in Symbolic Interaction* 5 (1984): 269–286.

———. "Fiction and Social Science: A Methodological Inquiry." Manuscript, 1986.

———. "The Group as Significant Other: Strategies for Definition of the Self." Paper presented at the Annual Meetings of the Pacific Sociological Association, San Francisco, April 1980.

———. *Hip Capitalism*. Beverly Hills, Calif.: Sage, 1979.

———. "Jenny's World." Manuscript, 1985.

————. "Midwestern Stories." Manuscript, 1978.

————. *The Mirror Dance: Identity in a Women's Community.* Philadelphia: Temple University Press, 1983.

————. "Research and the Construction of a Text." *Studies in Symbolic Interaction* 2 (1979): 167–187.

Lane, Peter. *Ceramic Form: Design and Decoration.* New York: Rizolli, 1988.

LeGuin, Ursula K. *Dancing at the Edge of the World: Thoughts on Words, Women, Places.* New York: Grove Press, 1989.

————. "It Was a Dark and Stormy Night; or, Why Are We Huddled about the Campfire?" In *On Narrative,* ed. W.J.T. Mitchell, pp. 187–195. Chicago: University of Chicago Press, 1981.

Levine, Frederick J., and Robert Kraus. "Psychoanalytic Theories of the Self: Contrasting Approaches to the New Narcissism." In *The Book of the Self: Person, Pretext, and Process,* ed. Polly Young–Eisendrath and James A. Hall, pp. 306–330. New York: New York University Press, 1987.

Lichtenberg, Joseph, and Samuel Kaplan, eds. *Reflections on Self Psychology.* Hillsdale, N.J.: Analytic Press, 1983.

Linden, R. Ruth. *Making Stories, Making Selves: Writing Sociology after the Holocaust.* Columbus: Ohio State University Press, 1991.

Lisle, Laurie. *Portrait of an Artist: A Biography of Georgia O'Keeffe.* New York: Washington Square Press, 1980.

Loevinger, Jane. "The Concept of Self or Ego." In *The Book of the Self: Person, Pretext, and Process,* ed. Polly Young-Eisendrath and James A. Hall, pp. 88–94. New York: New York University Press, 1987.

McCloskey, Donald N. *The Rhetoric of Economics.* Madison: University of Wisconsin Press, 1985.

McLuhan, T. C. *Dream Tracks: The Railroad and the American Indian, 1890–1930.* New York: Harry N. Abrams, 1985.

Mahler, Margaret S., Fred Pine, and Anni Bergman. *The Psychological Birth of the Human Infant: Symbiosis and Individuation.* New York: Basic Books, 1975.

Mairs, Nancy. *Plaintext: Deciphering a Woman's Life.* New York: Harper and Row, 1986.

Marcus, George E., and Michael M. J. Fischer. *Anthropology as Cultural Critique: An Experimental Moment in the Human Sciences.* Chicago: University of Chicago Press, 1986.

Marriott, Alice. *María: The Potter of San Ildefonso.* Norman: University of Oklahoma Press, 1948.

Mascia-Lees, Francis E., Patricia Sharpe, and Colleen Ballerino Cohen. "The Postmodernist Turn in Anthropology: Cautions from a Feminist Perspective." *Signs* 15:1 (1989): 7–33.

Mauss, Marcel. "A Category of the Human Spirit." *Psychoanalytic Review* 55:3 (1968 [1938]): 457–481.

Mead, George H. *Mind, Self, and Society.* Chicago: University of Chicago Press, 1962.

Messenger, Lisa Mintz. *Georgia O'Keeffe.* New York: Metropolitan Museum of Art, 1988.

Middlebrook, Diane. "The Life of a Good Old Girl." *Hudson Review* 41:3 (1988): 581–585.

Miller, Nancy K. "Changing the Subject: Authorship, Writing, and the Reader." In *Feminist Studies: Critical Studies,* ed. Teresa de Lauretis, pp. 102–120. Bloomington: Indiana University Press, 1986.

Millman, Marcia, and Rosabeth Moss Kanter, eds. *Another Voice: Feminist Perspectives on Social Life and Social Science.* New York: Anchor, 1975.

Minge, Ward Alan. *Acomá: Pueblo in the Sky.* Albuquerque: University of New Mexico Press, 1976.

Modell, Judith Schachter. *Ruth Benedict: Patterns of a Life.* Philadelphia: University of Pennsylvania Press, 1983.

Myerhoff, Barbara. *Number Our Days.* New York: Simon and Schuster, 1980.

Nelson, John S., Allan Megill, and Donald N. McCloskey, eds. *The Rhetoric of the Human Sciences: Language and Argument in Scholarship and Public Affairs.* Madison: University of Wisconsin Press, 1987.

Nisbet, Robert. *Sociology as an Art Form.* New York: Oxford University Press, 1976.

Oakley, Ann. "Interviewing Women: A Contradiction in Terms." In *Doing Feminist Research,* ed. Helen Roberts, pp. 30–61. London: Routledge and Kegan Paul, 1981.

O'Keeffe, Georgia. *Georgia O'Keeffe.* New York: Viking, 1977.

———. *Georgia O'Keeffe: Art and Letters.* Comp. Jack Cowart, Juan Hamilton, and Sarah Greenough. New York: New York Graphic Society Books, 1987.

———. *Some Memories of Drawings.* Ed. Doris Bry. Albuquerque: University of New Mexico Press, 1988.

Paget, Marianne A. "Experience and Knowledge." *Human Studies* 6 (1983): 67–90.

———. "Life Mirrors Work Mirrors Text Mirrors Life . . ." *Social Problems* 37:2 (1990): 137–148.

Peterson, Susan. *The Living Tradition of Maria Martinez.* Tokyo: Kodansha, 1978.

———. *Lucy M. Lewis: American Indian Potter.* Tokyo: Kodansha, 1984.

Polkinghorne, Donald E. *Narrative Knowing and the Human Sciences.* Albany: State University of New York Press, 1988.

Pollitzer, Anita. *A Woman on Paper: Georgia O'Keeffe.* New York: Touchstone, 1988.

Pratt, Mary Louise. "Fieldwork in Common Places." In *Writing Culture: The Poetics and Politics of Ethnography,* ed. James Clifford and George E. Marcus, pp. 27–50. Berkeley: University of California Press, 1986.

Price, Sally. *Primitive Art in Civilized Places.* Chicago: University of Chicago Press, 1989.

Rabinow, Paul. " 'Facts Are a Word of God': An Essay Review of James

Clifford's *Person and Myth: Maurice Leenhardt in the Melanesian World."* In *Observers Observed: Essays on Ethnographic Fieldwork,* ed. George W. Stocking Jr., pp. 196–207. Madison: University of Wisconsin Press, 1983.

———. *Reflections on Fieldwork in Morocco.* Berkeley: University of California Press, 1977.

Redfearn, Joseph W. T. "Terminology of Ego and Self: From Freud(ians) to Jung(ians)." In *The Book of the Self: Person, Pretext, and Process,* ed. Polly Young-Eisendrath and James A. Hall, pp. 384–403. New York: New York University Press, 1987.

Reinharz, Shulamit. *On Becoming a Social Scientist: From Survey Research and Participant Observation to Experiential Analysis.* New Brunswick, N.J.: Transaction Publications, 1988.

Reiter, Rayna R., ed. *Toward an Anthropology of Women.* New York: Monthly Review Press, 1975.

Richardson, Laurel. "The Collective Story: Postmodernism and the Writing of Sociology." *Sociological Focus* 21:3 (1988): 199–208.

———. "Narrative and Sociology." *Journal of Contemporary Ethnography* 19:1 (1990): 116–135.

———. *Writing Strategies: Reaching Diverse Audiences.* Newbury Park, Calif.: Sage, 1990.

Roberts, Helen, ed. *Doing Feminist Research.* London: Routledge and Kegan Paul, 1981.

Robinson, Roxanna. *Georgia O'Keeffe: A Life.* New York: Harper and Row, 1989.

Rosaldo, Michelle Zimbalist. "The Use and Abuse of Anthropology: Reflections on Feminism and Cross-Cultural Understanding." *Signs* 5:3 (1980): 389–417.

Rosaldo, Michelle Zimbalist, and Louise Lamphere, eds. *Woman, Culture, and Society.* Stanford: Stanford University Press, 1974.

Rosaldo, Renato. *Culture and Truth: The Remaking of Social Analysis.* Boston: Beacon, 1989.

Schafer, Roy. "Narration in the Psychoanalytic Dialogue." In *On Narrative,* ed. W.J.T. Mitchell, pp. 25–50. Chicago: University of Chicago Press, 1981.

Shils, Edward. *Tradition.* Chicago: University of Chicago Press, 1981.

Smith, Dorothy E. "A Sociology for Women." In *The Prism of Sex: Essays in the Sociology of Knowledge,* ed. Julia A. Sherman and Evelyn Torton Beck, pp. 135–187. Madison: University of Wisconsin Press, 1979.

———. "Women's Perspective as a Radical Critique of Sociology." *Sociological Inquiry* 44 (1974): 7–13.

Spence, Donald P. *Narrative Truth and Historical Truth: Meaning and Interpretation in Psychoanalysis.* New York: Norton, 1982.

———. "Turning Happenings into Meanings: The Central Role of the Self." In *The Book of the Self: Person, Pretext, and Process,* ed. Polly Young-Eisendrath and James A. Hall, pp. 131–150. New York: New York University Press, 1987.

Bibliography

Stacey, Judith. *Brave New Families: Stories of Domestic Upheaval in Late Twentieth Century America*. New York: Basic Books, 1990.

―――. "Can There Be a Feminist Ethnography?" *Women's Studies International Forum* 11:1 (1988): 21–27.

Stacey, Judith, and Barrie Thorne. "The Missing Feminist Revolution in Sociology." *Social Problems* 32:4 (1985): 301–316.

Strathern, Marilyn. "An Awkward Relationship: The Case of Feminism and Anthropology." *Signs* 12:2 (1987): 276–292.

Sullivan, Harry Stack. *The Psychiatric Interview*. New York: Norton, 1970.

Taylor, Charles. *Sources of the Self: The Making of the Modern Identity*. Cambridge: Harvard University Press, 1989.

Thorne, Barrie. "Political Activist as Participant Observer: Conflicts of Commitment in a Study of the Draft Resistance Movement of the 1960s." In *Contemporary Field Research: A Collection of Readings*, ed. Robert M. Emerson, pp. 216–234. Boston: Little, Brown, 1983.

Toulouse, Betty. *Pueblo Pottery of the New Mexico Indians*. Santa Fe: Museum of New Mexico Press, 1977.

Trimble, Stephen. "Talking With Clay." *New Mexico Magazine* 64:8 (August 1986): 41–47.

―――. *Talking with the Clay: The Art of Pueblo Pottery*. Santa Fe: School of American Research Press, 1987.

Turner, Victor W., and Edward M. Bruner, eds. *The Anthropology of Experience*. Urbana: University of Illinois Press, 1986.

Van Maanen, John, ed. "Special Issue: The Presentation of Ethnographic Research." *Journal of Contemporary Ethnography* 19:1 (1990).

―――. *Tales of the Field: On Writing Ethnography*. Chicago: University of Chicago Press, 1988.

―――. "Trade Secrets: On Writing Ethnography." Paper presented to the conference on Writing the Social Text, University of Maryland—College Park, November 1989.

Wall, Cheryl A. *Changing Our Own Words: Essays on Criticism, Theory, and Writing by Black Women*. New Brunswick, N.J.: Rutgers University Press, 1989.

White, Hayden. *Metahistory: The Historical Imagination in Nineteenth-Century Europe*. Baltimore: Johns Hopkins University Press, 1973.

―――. *Tropics of Discourse: Essays in Cultural Criticism*. Baltimore: Johns Hopkins University Press, 1978.

Williams, Patricia J. "On Being an Object of Property." *Signs* 14:4 (1988): 5–24.

Winnicott, D. W. *Playing and Reality*. New York: Basic Books, 1971.

WNET/Thirteen. *Georgia O'Keeffe*. Videotape. Boston: Home Vision, 1977.

Woolf, Virginia. *The Captain's Death Bed and Other Essays*. New York: Harcourt Brace Jovanovich, 1978.

Young-Eisendrath, Polly. "The Female Person and How We Talk about Her." In *Feminist Thought and the Structure of Knowledge*, ed. Mary McCanney Gergen, pp. 152–172. New York: New York University Press, 1988.